벽돌아, 벽돌아! 넌 뭐가 되고 싶니?
Brick, Brick! What do you want to be?

초판 발행	2018년 10월 15일
엮은이	담디 편집부 엮음
펴낸이	서경원
편집	나진연
디자인	이철주
펴낸곳	도서출판 담디
등록일	2002년 9월 16일
등록번호	제9-00102호
주소	서울시 강북구 삼각산로 79, 2층
전화	02-900-0652
팩스	02-900-0657
이메일	damdi_book@naver.com
홈페이지	www.damdi.co.kr

First Edition Published	October 2018
Compiler	DAMDI Publishing House
Publisher	Kyongwon Suh
Editor	Jinyoun Na
Art Director	Cheolju Lee
Publishing Office	DAMDI Publishing House
Address	2F, 79, Samgaksan-ro, Gangbuk-gu, Seoul, 01036, Korea
Tel	+82-2-900-0652
Fax	+82-2-900-0657
E-mail	damdi_book@naver.com
Homepage	www.damdi.co.kr

지은이와 출판사의 허락 없이 책 내용 및 사진, 드로잉 등의 무단 복제와 전재를 금합니다.

All rights are reserved. No part of this Publication may be reproduced, transmitted or stored in a retrieval system, photocopying, in any form or by any means, without permission in writing from DESIGNERS and DAMDI.

정가 25,000원

© 2018 DAMDI and DESIGNERS
   Printed in Korea
   ISBN 978-89-6801-084-2(93610)

이 도서의 국립중앙도서관 출판시도서목록(CIP)은 서지정보유통지원시스템 홈페이지(http://seoji.nl.go.kr)와
국가자료공동목록시스템(http://www.nl.go.kr/kolisnet)에서 이용하실 수 있습니다.(CIP제어번호: CIP2018029889)

Brick Brick
What do you want to be?

벽돌아, 벽돌아! 넌 뭐가 되고 싶니?

edited by 나진연(JinyounNa)

담디
DAMDI

Contents

006 **Essay**

    008 **Material Time** - Cheungvogl

    020 **Paradigm Shift** The Impossibility of Contemporary Architecture - SUPA architects schweitzer song

032 **How to use Material?** (Brick&Tile, Wood, Glass)
Interview with Architects

    034 Arenas Basabe Palacios Arquitectos
    048 ARPHENOTYPE
    070 AZC
    088 BOARD
    109 Carlos Lampreia
    118 Casanova+Hernandez Architects
    146 CEBRA
    158 Davide Macullo Architects
    172 Donner Sorcinelli Architecture
    181 Katsutoshi Sasaki+Associates
    187 Keiichi Hayashi Architect
    195 KimuraMatsumoto Architects office
    204 LANDÍNEZ+REY architects
    216 M artı D Mimarlık
    223 modostudio
    232 Mork-Ulnes Architects

241	murmuro
252	NISHIZAWA ARCHITECTS
264	Object-e architecture
283	OFIS arhitekti
288	OOIIO Architecture
303	Piuarch
312	SLOT STUDIO
324	SMAR Architecture Studio
332	Stefano Corbo Studio
342	stpmj
352	Studio Farris Architects
358	SUPA architects schweitzer song
367	TAKK Architecture
376	TheeAe Architects
382	TOUCH Architect
403	UNStudio

Essay

"IN OUR WORK THE MATERIALITY OF ARCHITECTURE IS THOUGHT TO BE A MEDIATOR BETWEEN HUMAN, ENVIRONMENT, MIND AND TIME, TO EVENTUALLY OVERCOME MATERIALITY."

- CHEUNGVOGL

008 **Material Time** - Cheungvogl

020 **Paradigm Shift** The Impossibility of Contemporary Architecture
- SUPA architects schweitzer song

Material Time
CHEUNGVOGL

물질적시간
CHEUNGVOGL

In our work the materiality of architecture is thought to be a mediator between human, environment, mind and time, to eventually overcome materiality.

우리는 건축의 물질성을 인간, 환경, 마음과 시간 사이의 중재자로 생각하며 궁극적으로는 물질성을 극복하려 한다.

Material in architecture is generally associated with its individual distinctive main characteristic. Concrete, for example, is often referred to as "cold" and "hard", whereas timber is described as a "warm" material. Steel, glass and aluminium are commonly associated with modernity, technology and seen as "cold". Metals commonly are seen as related to technology and felt as cold. Copper, in contrary, is often described as "warm", which might result from its coloration and also its softness. Generally, the reduced description material often undermines the ambiguity and true meaning of materiality. Architectural material is often employed based on or with fundamental focus on expressing its preconceived identity, leaving architectural intervention largely pre-defined by surfaces, with space and material devaluated of their opportunities

건축의 재료는 일반적으로 각 재료의 개별적인 주요 특성과 연관된다. 예를 들어, 콘크리트는 종종 "차갑고" "단단한" 것으로 불리는 반면 목재는 "따뜻한" 재료로 묘사된다. 강철, 유리 및 알루미늄은 일반적으로 근대성, 기술과 연관되며 "차가운" 것으로 간주한다. 금속도 일반적으로 기술과 연관되며 차갑게 느껴진다. 반대로 구리는 종종 착색과 부드러움 때문인지 "따뜻한" 것으로 묘사된다. 일반적으로, 이렇게 단면적인 묘사는 많은 경우 물질성의 다의성과 진정한 의미를 훼손한다. 건축 재료는 자주 이런 선입견에 기반을 두거나 이를 표현하는데 근본적인 초점을 두어 사용되기 때문에 표면적으로 미리 정해진 건축 재료의 이미지만이 남는다. 이는 공간과 재료의 기회와 가능성의 가치를 줄인다.

and possibilities.

To reflect upon the general affiliations of materials, as an example, we comprehend concrete to be a very tactile material in its haptic sensation and appearance based on its production methodologies, for instance with timber formwork construction. It can appear soft in form due to its mouldability, principally as concrete in construction state is a soft material and only hardens post production. Concrete as thermal mass can also be utilized to regulate and enhance room temperature sustainably and in this sense, the material literally performs as a "warm" material. If applied, produced and planned differently, concrete can be hard, cold and austere, yet these are not the only characteristics which solely describe the materials ambiguous and diverse qualities.

In the works of Peter Zumthor, Tadao Ando and Zaha Hadid, concrete unfolds different qualities, entirely expressing the diversified attributes of the material, which can sometimes be contradictory in their singular description. Nonetheless, beyond the physical definition, in each architect's work, the spatial and emotional

재료의 일반적인 연관성을 생각해보자. 예를 들어 우리는 콘크리트를 목제 거푸집 건설과 같은 생산과정에 기반한 촉감과 외관때문에 매우 촉각적인 재료로 생각한다. 게다가 쉽게 주조할 수 있어서 형태가 부드럽게 보일 수 있다. 건설 중의 콘크리트는 대게 유연하며 타설 후에 굳는다. 열 질량으로서 콘크리트는 실온을 지속가능하게 조절하고 향상시키는 데 사용할 수 있으며, 이러한 의미에서 콘크리트는 문자 그대로 "따뜻한" 재료가 된다. 다르게 적용, 생산 및 계획하면 콘크리트는 단단하고 차갑고 엄격할 수 있지만, 이 재료의 다의적이고 다양한 특성을 설명하는 유일한 특성은 아니다.

피터 줌터(Peter Zumthor), 안도 다다오(Tadao Ando) 및 자하 하디드(Zaha Hadid)의 작품에서 콘크리트는 다양한 특성을 드러낸다. 이들은 때로 단순한 설명에서 모순될 수 있는 이 재료들의 다양한 특성을 완전히 표현한다. 어쨌든 물리적 정의를 넘어, 각 건축가의 작업에서 같은 재료로 이루어진 공간적 및

narrative achieved with the same material is distinctively particular with unique interpretations resulting from similar materiality.

In common language, architectural materials have widely become synonymous for social, economic and urban deficits, separating the material from its materiality and transforming it into a political and critical projection. The term "concrete jungle" for example describes the overgrowing urban context, which is seen as anonymous and rejecting to human life, whereas steel and glass towers are conceived as cold and symbols of power and financial supremacy. As appropriate as the criticism of urban and coherently social and economic deficiencies may be, these references do not reflect the character and nature of the material itself. In this way, the material has become the generalized symbolic replacement for its misuse, a stigma to its reputation, as architecture and its employment of materiality has in fact always been subject to expression of political and economic intend since the beginning of civilisation.

In a natural environment, the employment

정서적 내러티브는 유사한 재료안에서의 독특한 해석이며 서로 아주 다르다.

실생활에서 건축 자재는 사회적, 경제적 및 도시적 결손의 동의어가 되어버렸다. 이는 재료를 물질성과 분리하고 정치적, 비판적 투영으로 변형한다. 예를 들어, "콘크리트 정글"이라는 용어는 익명으로 인간의 삶을 거부하고 과도하게 성장하는 도시 상황을 묘사하는 반면, 강철과 유리 타워는 차가운 권력과 재정적 패권의 상징으로 인식된다. 도시와 일관된 사회 경제적 결함에 대한 비판이 적절할 수 있지만, 이러한 설명은 물질 자체의 성격과 본질을 반영하지 않는다. 이런 식으로 재료는 오용으로부터 일반화된 상징성을 얻음으로써 재료 자체의 명성에 오명이 되어버렸다. 문명의 시작부터 건축과 물질성은 사실 항상 정치 및 경제적 의도를 표현했어야 했다.

지연환경에서 건축 자재는 명백

of architectural materials is evidently related to local resources, landscape and topography, creating a relationship between inside and outside, individual, community and the surrounding nature. In the urban context, the relationship between natural environment, recourses and human life appear rather abstract and even unrelated. This abstraction can lead to the perception of the build environment and architecture as cold, isolating and inhuman, with material becoming a physical and psychological separator of all factors and parameters.

In Cheungvogl's work, we are foremost interested in the consideration of humanitarian values. Material becomes subject to physiological and emotional dematerialization, bringing humans to the centre point of design thinking rather than ornamental materiality. In this way we see material as the supporter and transporter, serving a "greater cause", and not only for the means of architecture itself. As architecture evolves around human life and is built upon human needs in correlation with its environment, architectural material is essentially a mediator between humans, environment and nature. In dense city context,

하게 지역 자원, 경관 및 지형과 연관이 있고 건물 내부 및 외부, 개인, 지역 사회와 주변 환경 간에 관계를 형성한다. 도시적 맥락에서 보면 자연, 의지 및 인간 사이의 관계는 다소 추상적이거나 심지어 관련이 없는 것처럼 보인다. 이러한 추상화는 건축 환경과 건축을 냉담하고 고립되고 비인간적인 것으로 인식하게 하며, 물질은 모든 요인과 특성의 물리적, 심리적 분리 요인이 된다.

Cheungvogl의 작품에서 우리는 인도주의적 가치에 가장 관심이 있다. 재료는 생리적 및 정서적 비물질화의 대상이 되어 장식적인 물질성보다는 인간을 디자인 사고의 중심점으로 다룬다. 이런 식으로 우리는 재료를 건축 자체만을 위한 수단이 아니라 "대의"를 위한 후원자이자 운송자로 본다. 건축은 인간의 삶을 중심으로 진화하고 환경과의 상호관계에 있는 인간의 필요에 따라 형성하므로, 건축 자재는 본질적으로 인간, 환경 및 자연 사이의 중재자이다. 고밀도의 도시 콘텍스트에서 건축 재료는 점차 자연환경을 대체하고 개인과 공공 공용공간에서의 인간의

architectural material gradually becomes a replacement of the natural environment and a mediator between individuals and their needs in a public communal space. Materiality in architecture is equally connected to anthropology as it is reliant on environmental sciences.

One material which overcomes the notion of separation naturally by its specific characteristics is glass. Its hard surface and physical attributes stand in stark contrast to its transparent and translucent appearance and quality, perceiving it a rather soft and sensual material. Its ability to draw natural light within enclosed spaces and the visual connection it provides between inside and outside allows us to employ our most favourite material, which is Time.

The redevelopment of the 110 year old department store Au Pont Rouge, within the protected UNSESCO World Heritage site of Saint Petersburg, Russia is based and thrives on the idea of mediation. The project mediates between the historic structure and the new architectural intervention, as well as it creates a relationship between inside and the outside, connecting the

요구를 중재하게 된다. 건축의 물질성은 환경 과학에 의존하기 때문에 인류학과도 똑같이 연결된다.

특별한 특성 덕분에 자연적으로 분리라는 개념을 극복하는 한가지 재료는 유리이다. 유리의 단단한 표면과 물리적 속성은 투명하고 반투명한 외관 및 특성과는 완전히 대조를 이루며 다소 부드럽고 관능적인 물질로 인식된다. 게다가 밀폐된 공간 내에 채광을 끌어들이는 능력과 내,외부의 시각적 연결 덕분에 우리가 가장 좋아하는 재료인 시간을 사용할 수 있다.

러시아 상트페테르부르크의 유네스코(UNSESCO) 세계 유산 보호 구역에 있는 110년 된 백화점 오 퐁 루즈(Au Pont Rouge)의 재개발은 중재에 기반을 두고 진행한다. 이 프로젝트는 역사적인 구조와 새로운 건축 인터벤션을 중재한다. 뿐만 아니라 내부와 외부 사이의 관계를 형성하여 백화점을 역사적으로 사회적 관련성이 있는 장소로서 주변에 연결한다.

department store as a place of historically social relevance to the public city life and context.

The semi-transparent glazing of the interior spaces is layered around the historic structure, without colliding or covering the original artefacts. While defining multifunctional galleries spatially, the translucent walls allow natural light to penetrate deep into the building and its historic atrium space and transform its appearance with changing light qualities throughout the day and seasons. The distinct light qualities of Saint Petersburg, varying from crisp blue tones in the cold season and in the morning hours, illuminate to warm orange tonalities during the afternoons and in the summer, transform the atmosphere of the interior space in accordance with the environment, while the semi-opaque layers open up to offer views into the historic city centre and over the adjacent Moika river. Through the translucent layers passers-by appear blurrier, the further they move away, making the closeness and distance of people experiential. To dissolve time and space coherency further, the industrial concrete floor, with concrete generally

내부 공간의 반투명한 유리는 기존 건물과 충돌하거나 덮어 버리지 않고 역사적인 구조물 주위에 층을 이룬다. 반투명한 벽은 다용도 갤러리를 공간적으로 정의하고, 빛이 건물과 역사적인 아트리움 공간 깊숙이 들어오게 하여 낮과 계절에 걸쳐 변화하는 빛의 특성으로 그 외관을 변화시킨다. 선명한 푸른 색조가 다양한 아침과 추운 계절부터 따뜻한 오렌지 색조가 밝히는 오후와 여름까지, 상트페테르부르크 빛의 뚜렷한 특징은 환경의 변화에 따라 실내 공간의 분위기를 변화시킨다. 반쯤 불투명한 유리 층을 열면 역사적인 도심과 인접한 모이카 강으로의 전망이 보인다. 반투명한 층을 통해 통행인은 멀리 이동할수록 흐릿하게 보이며 사람들의 근접성과 거리감을 하나의 경험으로 만든다. 일반적으로 단단함과 관련되는 콘크리트지만, 시간과 공간의 일관성을 더욱 해체하기 위해 여기서 콘크리트 바닥은 거의 수면같이 빛을 부드럽게 반사하여 공간의 채광을 향상시킨다.

014 Brick, Brick! What do you want to be?

Au Pont Rouge

associated with hardness, softly reflects the light environment, almost like a water surface, enhancing the natural light qualities of the space.

Within Au Pont Rouge the presence of time is omnipresent. The setting places the visitor at the centre between history and present as it relates the individual to time and existence over the course of the day and seasons. The materiality of Au Pont Rouge allows the building to create a connection between inside and outside as well as present time and history. With natural light being the perfect representation of time, the one parameter which definitely relates humans to each other, to their environment and context, time is the ultimate and universal measurement of existence, documented by history.

Taghaus, a gallery, housing a private collection in Dusseldorf, Germany, is conceived with similar materials as Au Pont Rouge, yet the key focus of the architecture is on the perception of three-dimensional art and its experiential qualities and therefore inwardly directed. Taghaus is designed as an enclosed sculpture garden to protect the exhibited

오 퐁 루즈 내에서 시간의 존재는 어디에서나 느낄 수있다. 건물의 배경은 방문객을 역사와 현재 사이의 중심에 배치한다. 하루와 계절에 걸쳐 개인을 시간 및 존재성과 연관시킨다. 오 퐁 루즈의 물질성은 건물 내부와 외부 사이뿐만 아니라 현재와 역사 사이에 연결고리를 만든다. 자연광은 인간을 개개인과 자연환경, 콘텍스트 사이를 연결하는 변수 중 하나인 시간의 완벽한 표현이며, 시간은 역사로 기록된 궁극적이고 보편적인 존재성의 측정법이다.

독일 뒤셀도르프에 개인 컬렉션을 소장하고 있는 갤러리인 타그하우스(Taghaus)는 오 퐁 루즈와 비슷한 재료로 고안되었지만, 이 건축의 핵심은 3차원적 예술과 그 경험적 특성에 대한 인식에 있으므로 내부를 향해 디자인되었다.

art pieces from the environment, while allowing natural light qualities throughout the building.

The building does not utilize any artificial lighting to illuminate the sculptures apart from the sole condition of changing natural light and the varying shadows to present and showcase the three-dimensional qualities and depths of the art exhibits. The translucent and semi-transparent external glazed panels of the enclosure and separation walls blur the boundaries of space, allowing visitors to engage with the art in a seemingly undefined surrounding, which is only connected to the outside world through the incidence of daylight and the softened indication of the context through the façade's translucent glazing. The highly polished reflective concrete floor dissolves the hard edges between inside and outside and immediate adjacent exhibition rooms with the notion of dematerialization of surfaces.

Within the exhibition spaces of Taghaus, time becomes a visible reading on the changing surfaces of the sculptures, expressing the multiple facets and characteristics of the arts in changing play

타그하우스는 건물 전체에 채광을 들이는 동시에 환경으로부터 전시된 예술 작품을 보호하기 위해 벽으로 에워싸인 조각 정원으로 설계되었다. 전시회의 3차원적 특성과 깊이를 보여주기 위해 채광과 그림자를 다양하게 바꿔야 하는 유일한 경우 외에 이 건물은 조각품에 인공조명을 사용하지 않는다. 인클로저와 분리벽의 투명 및 반투명한 외부 유리 패널은 공간의 경계를 흐려 언뜻 보기에 규정되지 않은 환경에서 방문객이 예술과 교류할 수 있게 한다. 이 공간은 채광과 파사드의 반투명한 유리를 통해 보이는 완화된 콘텍스트의 표시를 통해서만 바깥 세상과 연결된다. 고도로 폴리싱되고 반사하는 콘크리트 바닥은 표면의 비물질화라는 개념에서 내부, 외부 및 인접한 전시실 사이의 단단한 경계를 용해한다.

타그하우스의 전시 공간 내에서 시간은 조각품의 변화하는 표면 위에서 볼 수 있는 해석이 된다. 변화하는 빛과 그림자를 지닌 예술의 여러 측면과 특성을 표현하고 인간과 예술적 의도 사이의

Essay 017

018 Brick, Brick! What do you want to be?

of light and shadow, creating a situation of intimate encounter between human and artistic intend. The engagement with the arts is elevated to a meditative, self-reflecting personal inner discourse with the object and the artist, rather than a spectacular observation at a glimpse.

Taghaus is conceived as the antithesis to common exhibition and gallery spaces, where the presentation of the arts is generally statically staged in curated light setting within a white box space, turning the arts into artificial and abstract objects with interrupted personal, emotional and intellectual connection to the spectator. The common exhibition space is designed for the arts to be seen. Taghaus instead employs time and light for the exhibited art pieces to be experienced individually. The materiality of architecture becomes a mediator between human, environment, mind and time, to eventually dissolve the categorical meaning of materiality.

친밀한 만남을 만든다. 예술과의 교제는 잠깐 보고 마는 장판이라 기보다 물체 및 아티스트와 하는 명상적이고, 자기 반성적이며 개인적인 내적 담론으로 향상된다.

타그하우스는 일반적인 전시 및 갤러리 공간에 대한 대조로 시작했다. 예술은 일반적으로 하얀 직사각형 공간 내에서 계산된 조명을 써 정적으로 전시하므로 관중에게 예술을 개인적, 정서적, 지적 연결이 잘린 인위적이고 추상적인 대상으로 전환시킨다. 일반적인 전시 공간은 예술을 볼 수 있도록 설계되었다. 그와 달리 타그하우스는 시간과 빛을 사용하여 전시된 예술 작품을 개별적으로 경험할 수 있도록 설계되었다. 여기서 건축의 물질성은 최종적으로 물질성의 범주적 의미를 녹이기 위해 인간, 환경, 마음과 시간 사이의 중재자가 된다.

Paradigm Shift
The Impossibility of Contemporary Architecture
Ryul Song, Christian Schweitzer

Paradigm Shift
현대건축은 자신의 의미를 잃었다!
송률, 크리스티안 슈바이처

Designing and discussing architecture in Korea uses the same terminology as it is used in the global discourse. But looking closer, these terms have a very specific and far more radical meaning in the Korean context than their western implications. Put into other words, what we know has no relevance in Seoul, what we do has no relevance in Seoul, everything is sucked up and swallowed by this overwhelming organism. Seoul stretches what we think to know about architecture as building to a point where it almost dissolves itself into an abstract task, disconnected completely from any pre-defined terminology. Experiencing Seoul therefore can consequently only result in a fundamental re-thinking and re-defining of every step in the generation of architecture.

Context ... the Bestand (best to be translated with the existing) is the starting point for any building task. We are trained to analyse the existing, to interpret the existing, to add to the existing, to build upon the existing, to

현재 한국의 건축 디자인과 토론에서 사용되는 (전문)용어들은 글로벌 토론에서 사용되는 용어들과 동일하다. 그러나 자세히 들여다보면, 이 용어들은 서양에서의 의미보다 한국에서 훨씬 더 극단적이며 매우 특이한 의미를 갖고 있다. 우리 건축가들이 이미 알고 있는 건축적 지식과 행위들은 서울에서 큰 의미가 없으며, 서울이라는 이 압도적 유기체는 모든 것을 집어 삼키고 있다. 우리가 이제까지 건축물을 위하여 배운 건축에 대한 지식의 줄은 이 곳 상황에 맞추기 위하여 늘리고 늘려보다가 결국은 끊어져 추상 속으로 용해되었으며, 서울의 건축은 이미 정의된 건축용어들과 완전히 분리되어 있다. 그러므로 서울을 경험하게 되면, 건축 생성의 모든 과정을 근본적으로 다시 생각하고 재정의 하여야만 한다는 것을 알게 된다.

컨텍스트(Context) … '기존 상황'은 건축 짓기 과정의 시작점이다. 우리는 기존상황을 분석하고 해석하며, 기존상황에 무엇인가를 첨부하고, 또한 그것을 기반으로 하여 상황을 완성시키도록 교육받았다. 우리의 디자인이 위치하게 될 대지 바로 옆의 물리적인

complete the existing. The immediate existing as well as the larger context into which we place our design is the reference point which legitimates and judges our actions. We can rely on the existing to establish a communication that links our design into the city and thereby becomes part of the existing.

In Seoul the existing withdraws itself very radically from this communication. An ostensible heterogeneity where everything refuses to relate to anything, where everything is only focused on itself and its own advantage, on the second glance reveals itself as an all-embracing sameness. Everything is same. Seoul is the endless repetition of one building programmed to the behaviour of its current user. Therefore every building becomes exchangeable without leaving a gap after its replacement which reduces the notion of context ad absurdum.

Time … usually one can rely on time, from the idea of a building to its execution to the aging of a building, and its eventual renewal or destruction.

In Seoul time is not linear; time exists only highly compressed to a single point where everything happens simultaneously. The very idea for a building in Seoul

컨텍스트이든, 범 형이상학적 컨텍스트이든, 컨텍스트는 우리 행위의 정당성과 타당성을 가늠 할 수 있는 기준점이 된다. 전형적인 의미로써, 도시 안에서 우리의 디자인은 이웃 건물과 소통을 확립하게 되며, 그럼으로써 기존상황의 한 부분이 되어간다. 그러나 서울에서 이 기존상황은 그 자신을 이러한 소통에서 배제시킨다. 모든 것이 다른 모든 것으로부터 관계 맺기를 거부하며, 모든 것이 그 자신과 자신의 이익에만 집중되어 있다. 이렇듯 표면적으로 매우 이질혼합적 (heterogeneity)인 서울은 각자가 매우 특색을 갖출 것이라 생각된다. 그러나 다시 보게 되면, 모든 것이 비슷해 보이는 동일성이 지배하고 있음을 알게 된다. 모든 것이 똑같다. 서울은 그 시점의 사용자의 행태로 프로그램화 된 하나의 건물이 끊임없이 반복된 곳이다. 그러므로 모든 건물은 그것의 컨텍스트 안에서의 잃어 버리게 될 의미를 애석해 할 필요 없이 다른 것으로 교체될 수 있다.

시간(Time) … 일반적으로 한 건축물은 시간의 흐름과 함께 산다. 처음의 아이디어, 그리고 짓기, 건물의 사용 그리고 낡음… 그리고 그것의 재생 또는 철거로 이어진다. 그러나 서울에서의 시간은 이렇게 선(線)적이지 않다. 서울의 시간은 어느 한 점, 즉 모든 것이 동시 다발적으로 발생하는, 대단히

Essay 021

022 Brick, Brick! What do you want to be?

©Ryul Song

©Ryul Song

024 Brick, Brick! What do you want to be?

implements already its obsolescence. The speed with which the city is built and rebuilt makes terms like constancy or continuity obsolete. The architect has to provide a solution to a task before it is even phrased. The same contradiction expresses itself in the lifespan of a building. It is detached from the building itself. The value of a building in contrast to its site is so low that it has no influence on it very right to exist. In Seoul every building is a temporary structure, with all the consequences a temporary structure is subject to.

Function … the overstressed expression "form follows function" has become a dogma in architecture. As beauty has become a subjective term, we legitimize our designs by developing a building that reflects, interprets or utilizes its function or functionality in a form generating process. We can rely on a specific function to provide us with a unique solution for every task.
Seoul is essentially multifunctional. Function and program of a building can change at any time even before it's built. In Seoul the built environment has the ability to adapt everything and to abandon it in the next moment.

압축된 한 점에서만 존재한다. 서울에서는 건물에 대한 개략적인 생각조차도 그것이 시작됨과 동시에 벌써 진부해지기 시작한다. 도시를 형성하고 재생하는 엄청나게 빠른 속도는 도시의 항상성 또는 지속성과 같은 조건들을 쓸모없게 만들어 버리며, 서울의 건축가는 과업이 말로 표현되기도 전에 그 과업의 해결책을 공급하여야만 한다. 이러한 모순은 건물의 수명에서도 똑같이 드러난다. 건물수명은 건물 자체와 상관이 없다. 건축물의 가치는 그것이 놓여있는 땅의 가치에 비하여 너무 낮기 때문에, 그 건축물이 그 곳에 존재할 수 있는 권리에 아무런 영향도 끼칠 수 없다. 그래서 서울의 모든 건축물은, 가설 건축물이 아님에도 불구하고, 가설 건축물로 정의 될 수 있는 모든 요건들을 갖춘 가설 구조체로 지어지고 있다.

기능(Function) … "형태는 기능을 따른다"라는 지나치게 강조된 표현은 건축에서 괴리가 되었다. 미(美)는 주관적인 요건이 되었기 때문에, 기능 또는 프로세스에서 탄생한 형태의 기능성을 반영/이용하는 건축물을 발전시킴으로써 우리의 디자인을 정당화 하려한다. 일반적으로 이러한 정당성을 극대화하기 위하여 특정한 프로그램(기능)으로부터 그것을 위한 유일무이한 해결책을 이끌어 낸 건축으로 향한다.

Therefore Seoul is widely undetermined; functions shift and overlap and exchange disconnected from its built structures. This program-overlap causes the extinction of any specific function; everything has to have the ability to let everything happen anytime anywhere else it becomes an interference factor in the city.

Space ... we are trained to design space, to structure and organize space, to establish special relationships, within a building, between buildings, within a city. Therefore architects developed a sophisticated differentiation of space, from private so semi-private, to semi-public, to public space with highly complex relations and exchanges that determine the quality of space.
Space in Seoul is flat. Space is reduced to the standard thickness of a wall, the separation of inside-space and outside-space. No other form of space exists. And even this wall is highly utilized as carrier of information, or storage, or privacy-generator, which shifts it from a spatial to a functional element that not just separates but disconnects from the space on the other side. Seoul in its last consequence is a continuous inside-space

그러나 서울은 근본적으로 다기능적 도시이다. 건축물의 기능과 프로그램은 언제든지 바뀔 수 있으며, 심지어 건물이 지어지기도 전에 바뀌기도 한다. 서울에서 건축환경은 모든 것이 적용될 수 있음과 동시에, 또한 바로 폐기될 수 있는 가능성을 내재하고 있다. 서울은 정의될 수 없는 도시이다. 이 곳에서 건축물의 기능은, 건축물 그 자체와는 무관하게 없어지고, 중첩되고, 교체된다.

공간(Space) ··· 우리는 일반적으로 공간을 설계하고, 공간을 구축하며 구성하도록 교육받았다. 그리고 한 건축물 안에서, 또는 건축물들 사이에서, 또는 도시 안에서 특별한 관계가 형성되도록 교육 받았다. 그래서 건축가는 아주 섬세하게 공간의 차이성을 발전시켜 나간다. 예를 들면, 대단히 복합적인 (관계와 공간의 질을 결정하는) 교류 안에서, 개인공간에서 준개인공간으로, 준공공공간, 그리고 공공공간을 발전시킨다. 그러나 서울의 공간은 2차원적이다. 공간은 내부공간과 외부공간을 나누는 벽(壁)의 일반적 두께로 축소된다. 다른 공간 형태는 존재하지 않는다. 그러나 이 벽 자체도 정보운송 또는 창고로 이용되며, 벽은 공간적 요소에서 기능적 요소로 전환된다. 더구나 벽은 공간을 단지 나누기

House P Plan Collage ©Ryul Song

Essay 027

which doesn't allow any reference point on the outside to set oneself in relation to the city. Inside a building in Seoul one is anywhere and nowhere at the same time. Quality of space becomes not a question of relation between different spaces but an abstract idea of space in relation to itself.

Form ... today everything is possible as long as one finds a good reason for it. In Seoul one doesn't even need this reason any more. Form-generating methodologies developed for a specific task are applied in Seoul almost randomly and disconnected from the theory behind it. Form is not generated anymore, it is applied; a subtle but crucial distinction. Form as the expression of an idea is reduced to form as the expression of form. Therefore Seoul has overcome form before it even realized it itself. Form disconnected from content, idea, methodology, excludes itself from an architectural process. Form in Seoul just is, whatever the implications of 'just being' can be.

Material ... as the material or materiality of a building is the dominating factor in the perception of a building next

만 하는 것이 아니라, 다른 쪽에 있는 공간으로부터 이 쪽을 완전히 단절시키는 사적공간제조기 (privacy-generator)로 최대한 이용되고있다. 서울은, 결과적으로 내부공간의 연속인 도시이며, 도시와의 관계성을 찾을 수 있는 외부와의 어떠한 기준도 허락하지 않는다. 그렇기 때문에 서울에서 누군가 건물의 내부공간에 있다는 것은, 어느 공간에 있든 차이가 없다는 것을 의미 한다. 이 곳에서 공간의 가치에 대한 물음은, 서로 다른 공간들이 만들어내는 관계성을 의미하는 것이 아니며, 하나의 공간 그 자체가 만들어 낸 (조형적인) 추상적 이미지를 의미하는 것이다.

형태(Form) ··· "오늘날은 모든 것이 가능하다"라는 관용구는, 그것을 위한 근거가있다면 유효한 것이다. 그러나 서울에서는 그 이유조차 더 이상 필요 없다. 어떤 특정한 프로젝트를 위하여 발전된 형태구현 방법론은 서울에서는 거의 무작위로 차용되고 있으며, 그 형태를 구현하려했던 본래의 이론과도 아무 연관성이 없다. 형태는 더 이상 새로이 구현되지 않으며, 다른 건축물로 부터 차용되고 적용될 뿐이다. 미묘하지만 매우 결정적인 차이이다: 발상의 표현으로써 형태는, 형태의 표현을 위한 형태로 의미 축소가 되었다. 그러므로 서울은 자신이 깨닫기도 전에 벌써 형태

House P ©김희천

Essay 029

to its form it is closely connected with the question of representation. We are trained to express our ideas though the application of specific materials. Material 'talks', it is coded through a flexible value system that puts our design into a larger architectural context. How we use material says who we are.

In a multifunctional, non-determined environment the question of material becomes a purely logistic one. Material is used according to its availability, not even necessarily according to its price. Seoul undercut this architectural value system where suddenly material doesn't provide information anymore about the status and the representation of a building. Material is used as a necessity not as a choice. As a building in Seoul only exists in the present, questions of durability, sustainability, maintenance of material do not arise and a much wider pallet of materials can be applied. Therefore Seoul has expanded the vocabulary of materials that are used in building by far. A simple plastic foil has the ability in Seoul to last longer and to talk more than any established building material.

With these observations in mind very fundamental questions arise. How can we

라는 이슈를 극복하였다. 형태는 내용, 아이디어, 방법론 등과 상관이 없으며, 그 자신을 이러한 건축 프로세스에서 배제시킨다. 서울에서 형태는 '단지 여기 있다'라는, 즉 어떤 것이라도 될 수 있다는 함축의 의미이다.

재료(Material) … 건축물의 재료 또는 재료성은 형태와 함께 그 건축물을 인식하는 지배적인 요소로써, 건축물 표상이라는 관건과 매우 밀접하다. 우리는 특정한 재료를 적용하여 우리의 아이디어를 표현하도록 교육받았다. 재료는 우리에게 '말을 한다'. 건축재료는 그 자체의 유연한 가치분류체계를 통하여 코드화되며, 이러한 유연한 가치 분류체계는 우리의 디자인을 더 넓은 범위의 건축적 컨텍스트 안에 놓일 수 있도록 한다. 우리가 재료를 어떻게 쓰는가 하는 것은, 우리가 어떤 건축가인가를 말하는 것이다.

그러나 다기능적이며 비확정적 환경에서, 재료에 관한 문제는 완전히 효율적인 유통에 대한 것이다. 재료는 가용성에 따라 사용되며, 가격도 2차적인 문제이다. 재료가 더 이상 건축물의 현황이나 표상에 관한 정보를 제공하지 않는 서울은 이러한 건축적 가치 분류체계를 무기력하게 만든다. 건축재료는 적합한 상황을 위한 선택사항이 아닌 즉각적인 필요성에 의해 사용되고 있다. 서울에서

design a building that doesn't want to be anything, that doesn't want to relate to anything, that doesn't want to represent anything, a building that essentially has no function, no economic life-time, no value to anyone? How can we design space that doesn't want to be organized, structured, determined, that doesn't want to establish relationships within a building, between buildings, within a city. How can we still communicate about and through architecture with out-dated definitions in the back of our heads of what establishes it? Seoul today forces architecture to reinvent itself.

건축물은 오직 현재에만 존재한다. 내구성 또는 지속가능성, 재료의 유지/보수에 대한 문제 제기가 없으므로, 더 광범위한 재료 선택이 가능하다. 서울은 건물에 사용되는 재료의 다양한 어휘가 단연코 굉장히 확장되어있다. 서울에서는 간단한 플라스틱 천조차도 건축 재료로 사용할 수 있다. 이러한 재료는 보다 더 긴 수명을 유지하며, 다른 어떤 일반적인 건축재료보다 더 많은 것을 이야기 한다.

이러한 서울에 대한 관찰로부터 근본적인 질문을 하게 된다. 어떤 것도 되고 싶지 않은, 무엇과도 연관되고 싶지 않은, 무엇도 표현하고 싶어 하지 않는 건축물을 어떻게 디자인하여야 하는가? 본질적으로 기능(프로그램)도 없으며, 경제적인 수명도 상관없으며, 어느 누구에게도 가치가 없는 건축물을 어떻게 디자인 하여야 하는가? 조직되기를, 체계를 갖추기를, 그리고 정의되기를 원치 않는 공간을 어떻게 디자인 하여야 하는가? 건축물 그 자체 안에서, 건축물들 사이에서 그리고 도시 안에서 어떠한 공간적 관계 맺기를 원치 않는 공간을 어떻게 디자인하여야 하는가? 우리의 생각 깊은 곳에 존재하는, 우리의 건축을 형성시킨 정의들이 이제는 진부해졌음에도 불구하고, 이 정의들로 건축물에 관한 소통이 여전히 가능한 것인가? 오늘의 서울은 건축이 혁신하도록 요구한다.

034	Arenas Basabe Palacios Arquitectos
048	ARPHENOTYPE
070	AZC
088	BOARD
109	Carlos Lampreia
118	Casanova+Hernandez Architects
146	CEBRA
158	Davide Macullo Architects
172	Donner Sorcinelli Architecture
181	Katsutoshi Sasaki+Associates
187	Keiichi Hayashi Architect
195	KimuraMatsumoto Architects office
204	LANDÍNEZ+REY architects
216	M artı D Mimarlık
223	modostudio
232	Mork-Ulnes Architects
241	murmuro
252	NISHIZAWA ARCHITECTS
264	Object-e architecture
283	OFIS arhitekti
288	OOIIO Architecture
303	Piuarch
312	SLOT STUDIO
324	SMAR Architecture Studio
332	Stefano Corbo Studio
342	stpmj
352	Studio Farris Architects
358	SUPA architects schweitzer song
367	TAKK Architecture
376	TheeAe Architects
382	TOUCH Architect
403	UNStudio

How to use Material?
(Brick&Tile, Wood, Glass) Interview with Architects

Arenas Basabe
Palacios Arquitectos

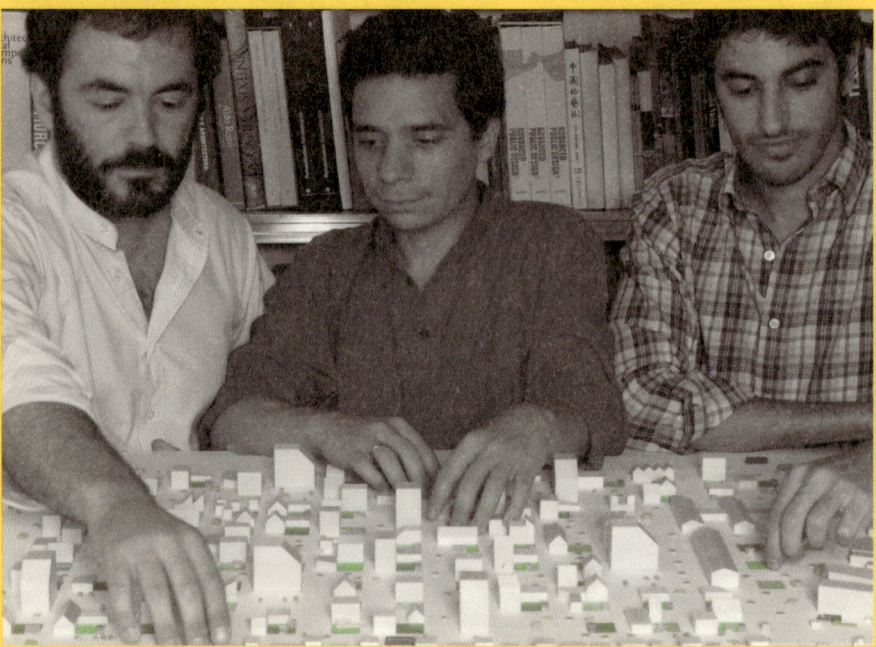

Who is...?

Arenas Basabe Palacios arquitectos is a young architecture and urbanism studio based in Madrid. Its partners Enrique Arenas Laorga, Luis Basabe Montalvo and Luis Palacios Labrador have been working together since 2006 and have won more than thirty prizes in architecture and urbanism competitions. They have given lectures in diverse institutions and presented their work and investigation in several international exhibitions. Their work has been published in Spain, France, Italy, Switzerland, UK, Austria, Germany, Cyprus, India and Korea.

Q1: What is material to an architect (or to you)?

A: At our practice aim to incorporate the use of materials to the design process, **understanding it as a tool that reinforces the underlying concept behind each project**. Associating a '**raw**' material, such as wood, stone or metal, with a structural architectural element of the project not only creates an element of reference to the users but also provides character to '**the built idea**'.

A: 우리는 **재료를 각 프로젝트의 기본 개념을 강화하는 도구**로 생각하고 디자인 과정에 통합하는 것이 목표이다. 목재, 석재 또는 금속과 같은 '**날 것**' 그대로의 재료를 프로젝트의 구조와 연관시키면 사용자가 판단 요소가 생길 뿐만 아니라 '**건설된 아이디어**'에 특성이 생긴다.

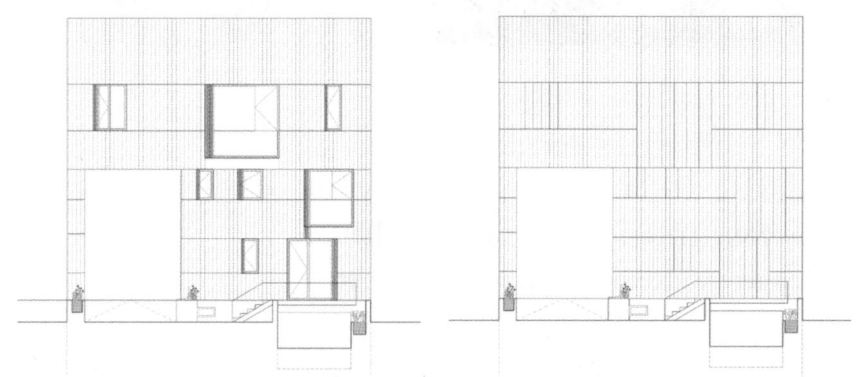

Box in the Box Elevation

Q2: Tell us about your favourite (or most often used) material and why.

A: In a market currently governed by the relationship between cost and aesthetics, we are under the threat of a rising demand of 'copy' materials: plastic

A: 예산과 미학의 관계에 지배되는 현재 시장에서는 나무와 유사한 플라스틱 마감재, 돌을 모방한 인쇄된 세라믹 타일 등 '복사'재료에 대한 수요

finishes that resemble wood, printed ceramic tiles that mimic stone, etc. In this regard, **we advocate the employment of 'raw' materials, such as stone, timber and metal, each one of them with their own inherent characteristics:** stiffness, maintenance requirements, and appearance to both eye and touch.

가 증가하고 있다. 그런 점에서 우리는 강도, 보수 사항 및 시각과 촉감과 같은 **각각 고유한 특성이 있는 돌, 목재 및 금속과 같은 본연의 재료 사용을 지지한다.**

Ninette House

Q3: When do you decide the material during the design process and what is your criteria? (e.g. budget, client's preference, design concept, climate, etc.)

From the initial stages of the design process the idea behind a project is deeply intertwined with a reflection about its materiality, which markedly determines its character, presence and intensity. We believe it is our duty to fulfill this underlying idea, which often entails overcoming the client's incomprehension, budget limitations or on-site challenges.

A: 디자인 과정의 초기 단계에서 프로젝트의 아이디어는 그 특성, 존재감 및 강도를 현저하게 결정하는 물질성에 대한 반영과 깊이 얽혀 있다. 이 근본적인 아이디어를 실현하는 것이 우리의 의무라고 믿는다. 이를 위해서 종종 의뢰인의 이해력, 예산 제한 또는 현장 문제를 극복해야 하기도 한다.

Q4: What are some architectural projects that inspired you regarding brick, tile, wood and/or glass? And why?

A: We are genuinely interested in architectural projects in which the use of materials is understood as part of a research process. Due to the enormous challenge that is to break with the conventionality of the predominant construction methods, any project that manages to redefine standardization parameters and manipulate the established conventions constitutes a source of stimulus, which encourages us to carry on working in pursuit of '**the built idea**'.

A: 우리는 재료의 사용법이 연구 과정의 일부인 건축 프로젝트에 진정으로 관심이 있다. 주요 건설 방식의 관습을 깨는 것은 엄청난 도전이기 때문에 표준화된 많은 변수를 새로 정의하고 확립된 규칙을 바꾸는 모든 프로젝트는 자극의 원천이 되어 우리가 '**건설된 아이디어**'를 추구하도록 북돋아 준다.

Q5: Tell us about the materials you are interested in or want to use in your projects right now.

A: At present we are working on a dual project, the '**Sister Houses**', two adjoining houses for two sisters and their families. In many ways the houses resemble their owners: despite certain common traits, each one of them has a specific character, which is expressed through the use of 'raw' materials. On the ground floor level, the

A: 현재 우리는 이중 프로젝트인 '**시스터 하우스** (Sister House)'를 진행하고 있다. 두 자매와 그 가족을 위한 두 개의 맞닿은 주택을 짓는 프로젝트이다. 여러 면에서 이 주택은 소유자와 닮았다. 일부 공통점에도 불구하고 각 주택에는 본연의 재료를 사용하여 표현한 독특한 특성이 있다.

We advocate the employment of 'raw' materials, each one of them with their own inherent characteristics.

각각 고유한 특성이 있는
본연의 재료 사용을 지지한다.

Box in the Box ©Imagen Subliminal

houses share a common plinth, a system of walls made of white board-formed concrete. However, each house employs wood in a unique way: in the first one the internal circulation core is clad with a continuous birch-tree panelling, whereas the facade of the second one is fitted with massive accoya window frames that seek to capture the landscape.

이 주택은 1층에 흰색 나무판자 무늬 콘크리트 벽으로 된 공통된 토대를 공유하지만 각 집은 나무를 다른 방식으로 사용한다. 첫 번째 집에서는 내부 동선 코어가 계속 이어지는 자작나무 패널로 덮인 반면, 두 번째 집의 파사드에는 풍경을 담기 위한 거대한 아코야 나무(Accoya) 창틀이 있다.

Sister House Project

TILE-Q1: Tell us about your favourite project that you used tile in or another architect's work - interior, facade, etc.

A: The project that best employs tiles in order to build an idea is 'Ninette House', a single family house located in the outskirts of Madrid. On the ground floor level, the flooring, a pixelated design made of micro-terrazzo tiles, extends beyond the glazing in a subtle color gradient, blurring the boundaries between the building and the surrounding landscape.

A: 아이디어 하나를 건설하는 데 타일을 가장 잘 사용한 우리 프로젝트는 마드리드 외곽에 있는 단독주택인 니네트 하우스(Ninette House)이다. 1층에 마이크로 테라초 타일로 된 픽셀 디자인의 바닥재가 미묘한 색상 그라디언트를 따라 유리를 건너서 건물과 주변 경관 사이의 경계를 흐린다.

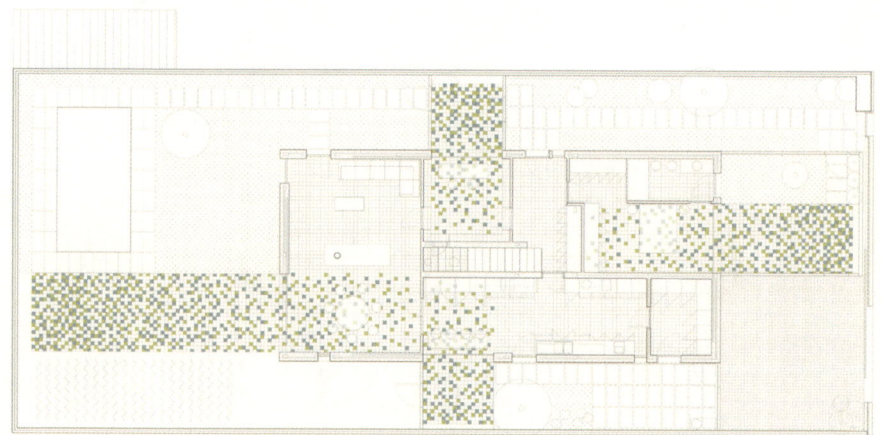

Ninette House-Ground floor plan

Arenas Basabe Palacios Arquitectos 041

01 COPING. 0.8MM THICK ZINC SHEET
02 WATERPROOF MEMBRANE
03 3MM THICK GALVANISED STEEL SHEET
04 BRICK WALL
13 80MM EXTRUDED POLYSTYRENE (EPS) INSULATION
14 15MM GYPSUM BOARD SUSPENDED CEILING
16 WATERPROOF MEMBRANE
17 20MM EXTERIOR PLASTER
18 GLASS FIBER SHEET
19 30MM THICK FLAT BRICK
21 BRICK WALL
22 20MM POLYURETHANE FOAM INSULATION
23 80MM POLURETHANE FOAM INSULATION
24 60MM CAVITY
25 BRICK WALL
26 20MM PLASTER
28 100X70MM STEEL PROFILE
29 3MM THICK GALVANISED STEEL SHEET
44 COMPACTED SOIL
61 30MM LEVELLING MORTAR
62 150MM CONCRETE DECK
63 50MM LEAN CONCRETE
64 120X120X30MM TERRAZZO
66 20MM PLASTER
67 70MM BRICK WALL
72 120X70MM TUBULAR STEEL PROFILE
73 3MM THICK GALVANISED STEEL SHEET WINDOW SILL PAN
74 CONCEALED ALUMINUM FRAME WINDOW
75 MOTORISED VENETIAN BLIND WITH ADJUSTABLE PROFILE SLATS
76 150X120MM STEEL PROFILE
77 CEMENT SCREED

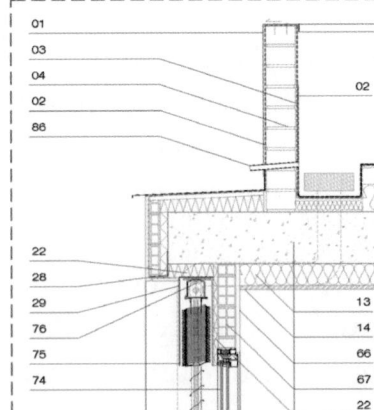
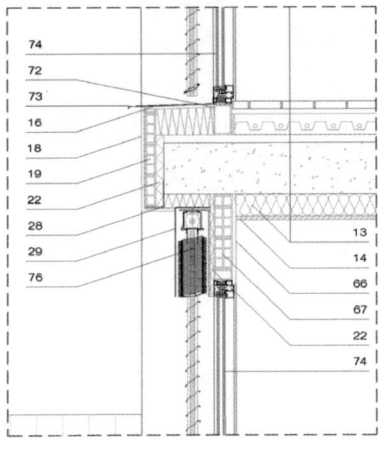
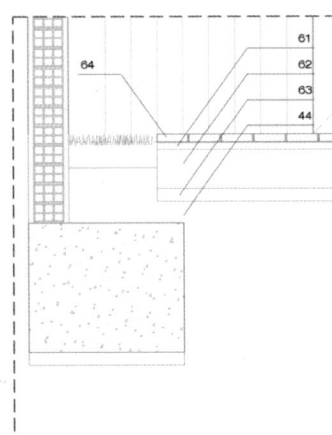
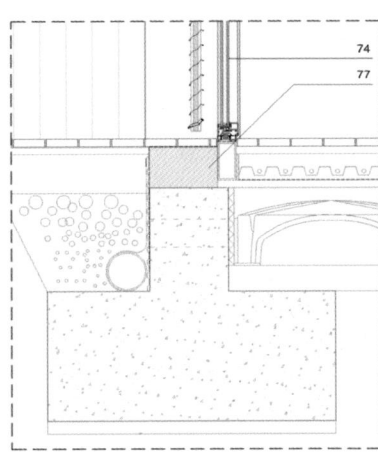

Ninette House Facede Detail

Ninette House ©Imagen Subliminal

TILE-Q2: What are the strengths and weaknesses of tile?

A: The versatility of the micro-terrazzo tiles allows us to use them both inside and outside, varying only in finish (either polished or rough). This allows us to reinforce the idea of a seamless continuity between interior and exterior spaces. The main 'weakness' of this material could be associated to the irregularity of its edges, which could stand in the way of the sharpest detailing. Nonetheless, we do not perceive it as a weakness rather than one of its inherent qualities. No tile is identical to another one, turning imperfection into one of the sources of its appeal.

A: 마이크로 테라초 타일은 그 유동성 덕분에 내부와 외부 모두에 사용할 수 있으며 마무리 (폴리싱 혹은 거침)만 다르다. 이는 내부와 외부 공간 사이의 원활한 연속성을 강화할 수 있게 해준다. 이 재료의 주 '단점'은 가장자리가 고르지 못한 점이며, 아주 깨끗한 디테일에는 방해가 될 수 있다. 하지만 우리는 이 점을 고유한 특성 중 하나로 생각하지 단점으로 보지 않는다. 완전히 같은 타일은 하나도 없으며 바로 이 불완전함이 타일의 매력 중 하나이다.

Terrazzo pavement Haifa ©Shlomo Katzav

WOOD-Q1: Tell us about your favourite project that you used wood in or another architect's work - interior, facade, etc.

A: We have explored the potential of wood to become an element of reference in the project for the renovation of a former industrial warehouse and its conversion into a center for a cultural association. Our 'Box in the box' accommodates the totality of servant spaces (circulation, foyer, toilets, storage and MEP services) within a massive wooden core made of OSB boards. The engineered lumber boards clad every surface of the core: floors, walls and ceilings. The employment of wood confers a recognizable character to this new component, boosting navigability and readability by users from both indoors and outdoors, as well as further consolidating the identity of the building as a cultural center.

A: 우리는 전 산업 창고의 개조 및 문화 협회 센터로 전환하는 프로젝트에서 목재의 잠재력이 판단 요소가 될 가능성을 탐구해보았다. 우리의 '상자 속의 상자' 프로젝트는 OSB 보드로 만든 거대한 나무 코어 안에 봉사하는 공간 (동선, 휴게실, 화장실, 저장 및 MEP 서비스)을 모두 수용한다. 공학목재로 바닥, 벽 및 천장과 같은 코어의 모든 표면을 덮었다. 목재로 이 새로운 구성에 알아보기 쉬운 특성을 부여하고, 실내 및 실외 사용자의 가항성과 가독성을 높이고, 문화 중심지로서의 건물의 정체성을 통합했다.

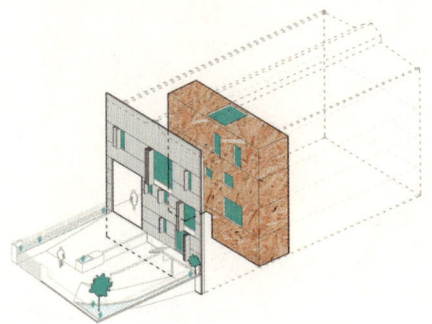
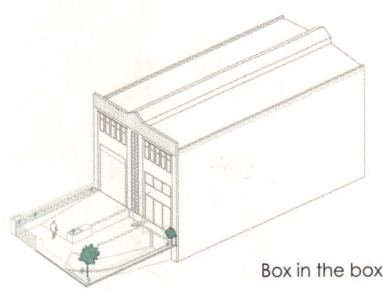

Box in the box

Arenas Basabe Palacios Arquitectos

WOOD-Q2: What are the strengths and weaknesses of wood?

A: The presence of wood in a project usually entails the existence of a strong duality between it and the rest of the architectural components. In our 'Box in the box' project, the forcefulness of the OSB boards that clad the servant core greatly contrasts with the clean lines and neutral-toned materials (grey continuous polished-concrete floors and a mostly white palette of translucent, transparent and opaque walls) employed in the main rooms: indoor playing court, assembly hall, chapel, classrooms, meeting room and offices. The idea of duality can also be acknowledged between the inner wooden box and the new metallic shell that wraps the existing facade, making the idea of a 'Box in the box' more meaningful.

A: 일반적으로 한 프로젝트에서 목재의 존재는 다른 건축 요소와 강한 이중성을 수반한다. 우리의 '상자 속의 상자' 프로젝트에서 봉사하는 공간을 감싸는 OSB 보드의 박력은 깨끗한 라인과 뉴트럴한 톤의 재료(회색 광택 콘크리트 바닥과 반투명, 투명, 그리고 불투명한 벽으로 된 흰색 팔레트)로 된 실내 경기장, 강당, 예배당, 교실, 회의실과 사무실 같은 주요 공간과 크게 대조된다. 이 이중성은 내부의 나무 상자와 기존 파사드를 감싸는 새로운 금속 껍질 사이에서도 볼 수 있어 '상자 안의 상자'에 대한 아이디어에 더 의미가 생긴다.

Box in the box ©Imagen Subliminal

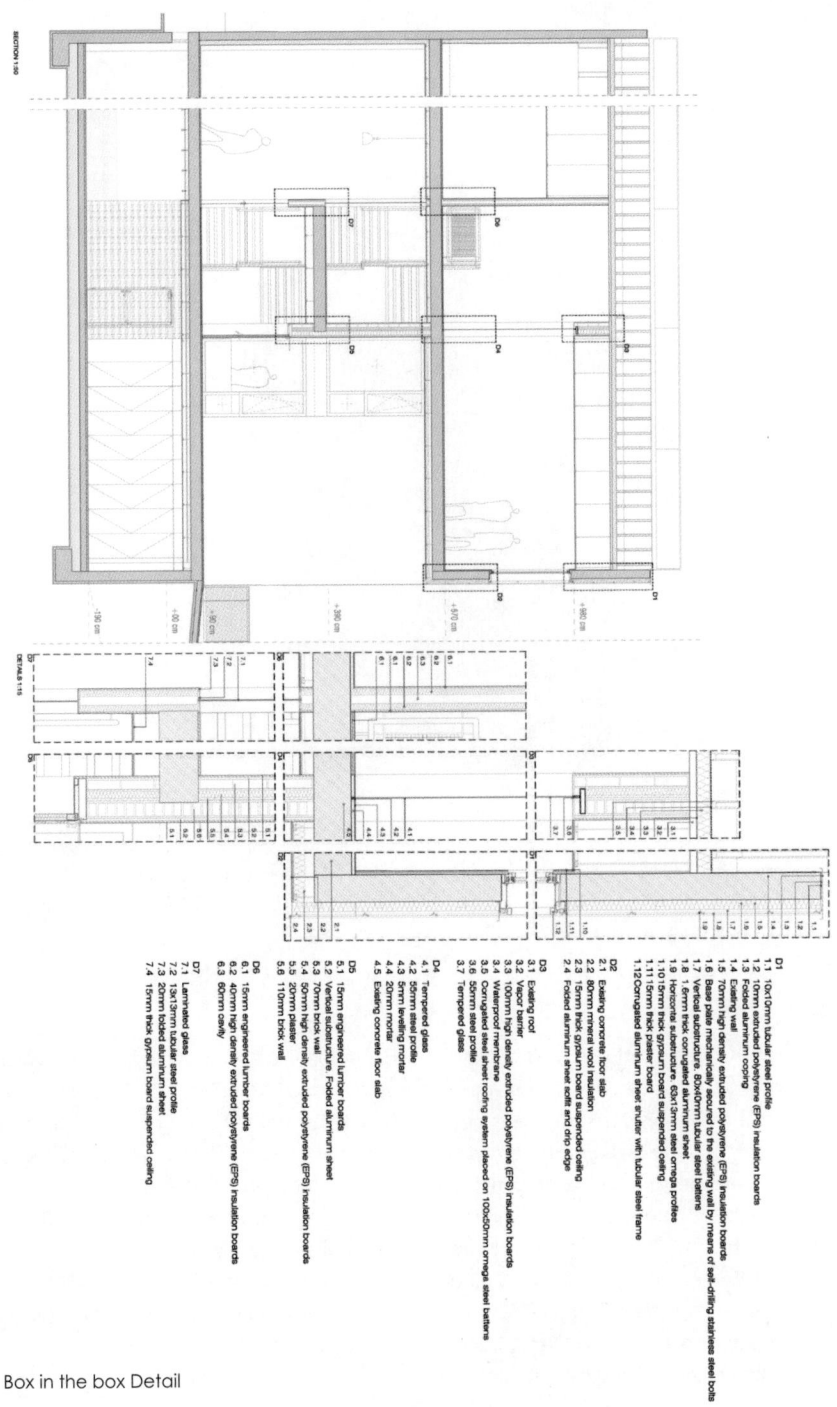

Box in the box Detail

Arenas Basabe Palacios Arquitectos

ARPHENOTYPE

Who is...?

Dietmar Köring, Dipl.-Ing.(FH) M.Arch. Architect BDA, is an architect, researcher, and educator living in Cologne. He is head of the architectural research office Arphenotype, where he focuses on blurring the boundaries of different artistic disciplines. From 2012 to 2017 he was a research fellow at TU Berlin / CHORA City & Energy and Dietmar has taught Digital Design at TU Braunschweig from 2010 to 2012, he was Guest Professor for Virtual Realities & Experimental Architecture at the University Innsbruck ./Studio3 in 2011, Technology and Design Lecturer at the Cologne Institute for Architectural Design / C-I-A-D and visiting lecturer for digital design at the DeMontfort University Leicester. From 2011 to 2012 he was assistant professor for Smart Grid research (Smart City Concepts 2022) at the Institute for Corporate Architecture at the Cologne Technical University.

Q1: What is material to an architect (or to you)?

A: **Material is like DNA, it makes the world we perceive.** Either natural, created by humans or based on new algorithms. We barely think about an object or a shape without materializing it. The world we perceive as human cannot exist without material.

Even the virtual world of games is based on data that I understand to be a material, or at least the intelligence of what constitutes a material. As data that needs to be mined, material can be mined. In addition, each material has its own data that determines behavior. Once we understand that, we can apply or combine material in a new way. We can speak about the "genetics" of material.

A: 재료는 DNA 같아서, 우리가 인식하는 세계를 만든다. 자연적이거나, 인공적이거나, 새로운 알고리즘을 기반으로 한다. 인간은 사물이나 모양을 구체화하지 않고는 거의 생각하지 않는다. 우리가 인간으로서 인식하는 세계는 물질 없이는 존재할 수 없다. 가상의 게임 세계조차도 내가 물질이라고 이해하는 데이터, 혹은 적어도 재료를 구성하는 정보에 기반한다. 데이터는 채굴해야 하고, 재료도 채굴할 수 있다. 또한 각 재료에는 성질을 결정하는 자체 데이터가 있다. 일단 이 점을 이해하면 우리는 새로운 방식으로 물질을 적용하거나 결합할 수 있다. 우리는 물질의 '유전학'에 대해 이야기할 수 있게 된다.

Q2: Tell us about your favourite (or most often used) material and why.

A: Because I work in an academic field, it is a material that stu-dents

A: 나는 학문 분야에서 일하기 때문에 내가 가장 좋아하는 재료는 학생들이 디지털 디

ARPHENOTYPE 049

use in digital design courses. The first time I saw this material in an architectural context was in a course led by Evan Douglis at the Si-Arc in 2007.

This material I am referring to is named Funky Foam. Funky Foam has excellent workmanship and is available in every color. It is very light, can be easily folded and glued. Students can use the laser cutter to create and cut out patterns and create three-dimensional space through folds or layers. In addition, it is very cheap, which leads to not cost limited prototypes, such as 3D prints that are still too expensive on a certain scale. What I also like is, that it has a certain thickness, which paper doesn't have. Certain shape studies can be outlined by needle points. However, I have to admit that it is a composite material that cannot be recycled and therefore is not a very sustainable material.

Q3: When do you decide the material during the design process and what is your criteria? (e.g. budget, client's preference, design concept, climate, etc.)

A: When we are working for clients or

자인 강좌에서 사용했던 재료이다. 건축적인 콘텍스트에서 이 재료를 처음 본 것은 2007년 SCI-Arc(Southern California Institute of Architecture - 남캘리포니아주 건축대)에서 에반 더글리(Evan Douglis)가 이끄는 강좌였다.

내가 언급하고 있는 이 재료는 "펑키 폼"이라고 불린다. 펑키 폼은 다루기 아주 쉬우며 색상이 아주 다양하다. 매우 가볍고, 쉽게 접히고, 붙일 수 있다. 학생들은 레이저 커터를 사용하여 패턴을 만들어 자르고, 접거나 층을 만들어 3차원 공간을 만들 수 있다. 또한 매우 저렴하여 특정 규모에서는 너무 비싼 3D 인쇄물처럼 제한된 프로토타입 이상을 만들 수 있다. 또 종이와 다르게 어느 정도 두께가 있다는 점도 좋다. 어떤 모양 연구는 바늘 끝으로 윤곽을 그릴 수도 있다. 하지만, 펑키 폼은 재활용 될 수 없는 복합재료이고 따라서 그다지 지속 가능한 재료는 아니라는 것을 인정한다.

A: 우리가 의뢰인이나 공모전

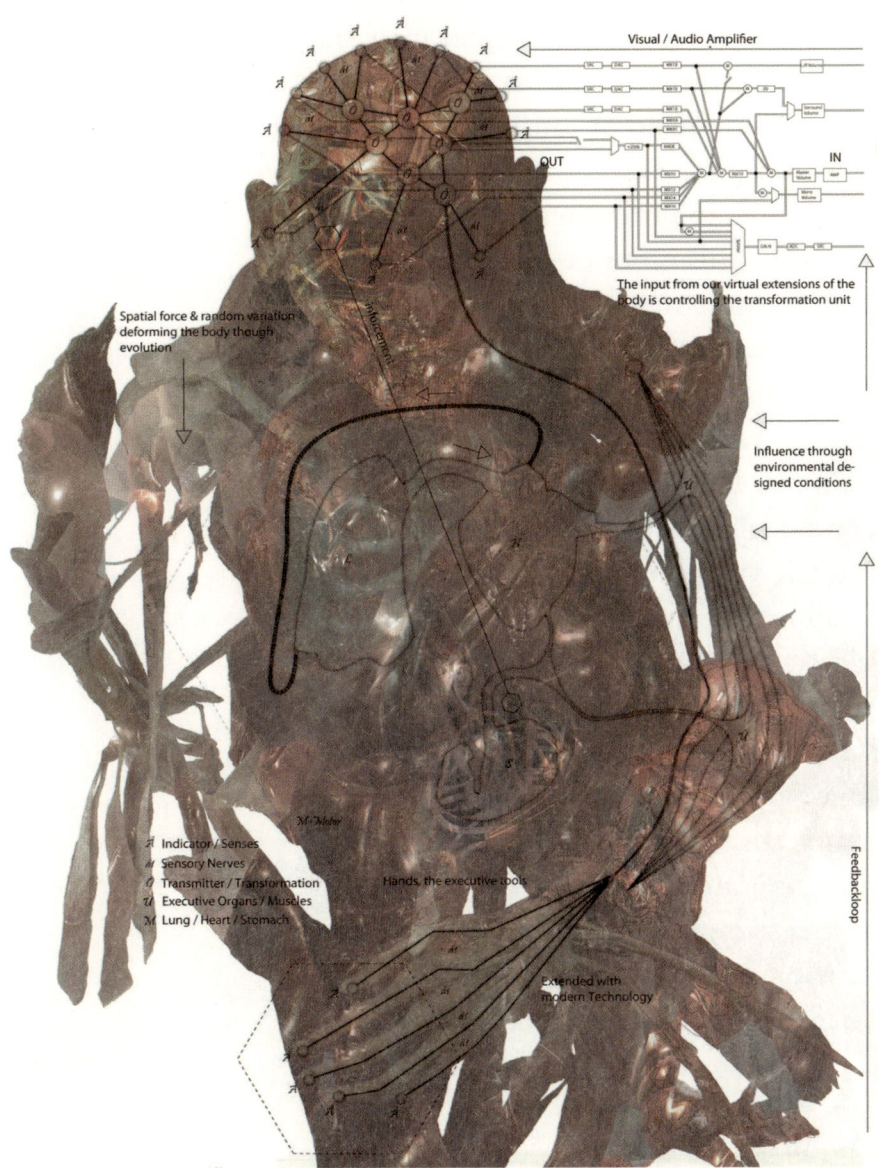

Material is like DNA.
재료는 DNA 같다.

competitions, it is mainly the budget, which determines the use of material. Often we would like to use more glass, but as glass is expensive, the use has to be well thought of. Then of cause also the brief. If we design a social housing complex, we would like to experiment with materials, but often this is not possible, as the client doesn't want to experiment. This is due to legal constraints (at least in Germany). However, we always try to evaluate the sustainable criteria of the material in its context. In 2007, we participated in a competition to design new high-voltage electricity pylons for Islands, where we proposed a construction of composite plastic - and since it was never used in such condition, we lost. Furthermore, we later discovered that there is a steel lobby producing these pylons, which in turn leads to the answer that a composite plastic is not desirable, even if it may be the future. In conclusion, **the choice of the material is mainly driven by the client's preference, budget and brief.**

을 위해 일할 때 재료를 주로 결정하는 것은 예산이다. 우리는 종종 유리를 더 많이 사용하고 싶지만 유리는 비싸기 때문에 사용할 때 아주 잘 생각해야 한다. 그리고 당연히 프로젝트 개요도 봐야한다. 우리는 공공주택 단지를 설계할 때 재료를 실험하고 싶지만, 의뢰인이 싫어하기 때문에 대부분 불가능하다. 이는 (적어도 독일에서는) 법적 제약 때문이다. 하지만 우리는 항상 그 콘텍스트에서 재료의 지속 가능함을 평가하려 한다. 2007년 우리는 한 섬에 새로운 고압 전기 철탑을 설계하는 공모전에 참가했다. 이 철탑을 복합 플라스틱으로 만들 것을 제안했는데, 그런 상황에서 한 번도 사용된 일이 없었기 때문에 공모전에서 졌다. 게다가 나중에 이런 철탑을 생산하는데 쓰일 강철을 위한 로비가 있다는 것을 알았다. 이로써 복합 플라스틱이 미래일지라도 강철보다 선호할 일이 없을 것이라는 답에 도달한다. 결론적으로, **재료의 선택은 주로 의뢰인의 선호도, 예산과 개요로 결정한다.**

Q4: What are some architectural projects that inspired you regarding brick, tile, wood and/or glass? And why?

A: When we look at Kondratieff's innovation cycles, I believe that we are in the prosperity phase for new materials. We develop new technologies that could lead to the development of new materials, or at least to the use of materials that have not been used in the architectural context. Neri Oxman, a Professor from the MIT, is researching coral growth patterns with the software Houdini. This might led to new forms, which a currently named "Quaddels". Hence material can be grown in shapes we want, as long as we control the behaviour of the material, the algorithm.

The architect and artist Tobias Klein from the City University of Hong Kong, has a similar approach, but with a different material: Tobias is applying the growth of crystals to sculptural form. Or the work of Rachel Amstrong, who is focusing on Proto-Cells, which would enable a material, which repairs itself. This are just three names in a new developing field of super-intelligent materials. Probably there are plenty of materials in our world, they communicate to us, they form what we see and perceive in our own reality. However, **in current time, we must think sustainable in the**

A: 콘드라티에프의 혁신 주기를 보면, 우리는 지금 새로운 재료의 번영 단계에 있다고 믿는다. 새로운 재료의 개발로 이어질 수 있는 새로운 기술을 개발하거나 적어도 건축적 맥락에서 사용되지 않은 재료를 사용한다. MIT의 교수인 네리 옥스맨(Neri Oxman)은 소프트웨어 후디니(Houdini)로 산호 성장 패턴을 연구하고 있다. 이것은 현재 "콰델(Quaddels)"이라 불리는 새로운 형태로 이어질 수 있다. 따라서 알고리즘으로 재료의 성질을 제어하는 한, 우리가 원하는 모양으로 재료를 키울 수 있다.

홍콩 시립 대학의 건축가이자 예술가인 토비아스 클라인(Tobias Klein)은 다른 재료로 비슷한 연구를 하고 있다. 토비아스는 결정의 성장을 조각적인 형태에 적용하고 있다. 또는 프로토셀(Proto-Cells)에 초점을 맞춘 레이첼 암스트롱(Rachel Amstrong)의 작업은 자체 수리가 가능한 재료를 만들 수 있을 것이다. 이는 초지능 재료라는 새로운 분야에서 단지 세 명의 이름이다. 아마도 우리 세계에는 많은 재료가 있고, 우리와 소통하고, 우리가 현실에서 보고 인식하는 것을 형성한다. 하지만 **오늘날 우리는 재료**

way, how we use materials – especially in architecture. Finally I would like to quote Buckminster Fuller, who said, that we must be responsible with materials, as they belong to all of us.

를 지속가능한 방법으로 사용하는 법을 생각해야 하며, 특히 건축에서 더 그러하다. 마지막으로 나는 벅민스터 풀러(Buckminster Fuller)의 말을 인용하고 싶다. 그는 재료는 우리 모두의 것이므로 책임감 있게 써야 한다고 말했다.

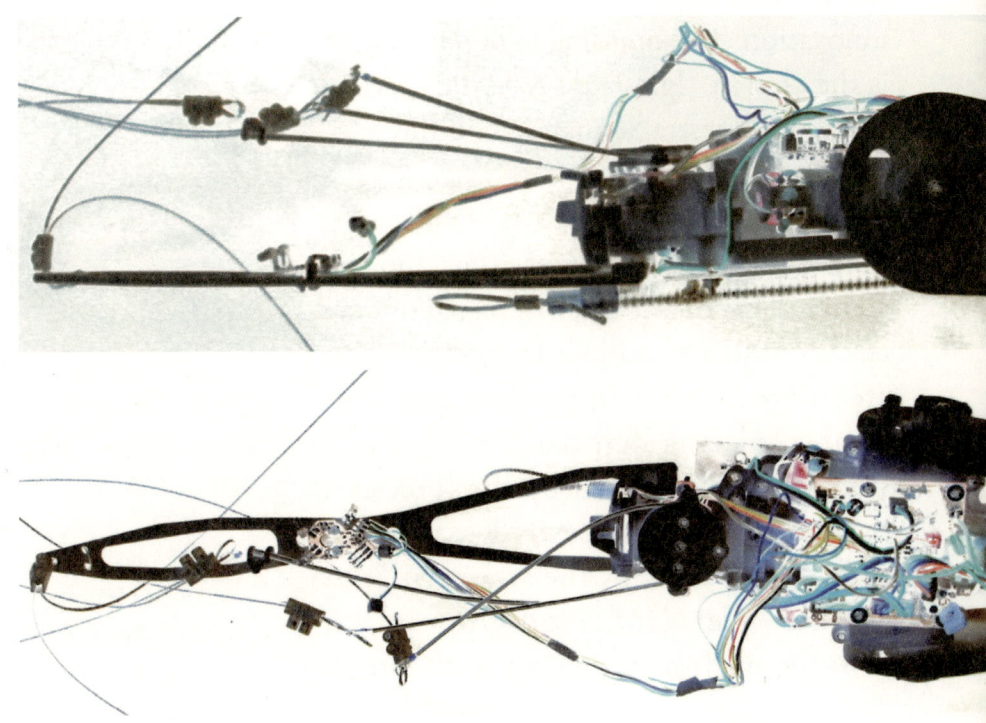

BRICK-Q1: Tell us about your favorite project that you used brick in or another architect's work - interior, facade, etc.

A: Bricks are an ancient material, which even today can be re-invented. One project which immediately pops up in my mind is the Kolumba Museum in Cologne by Peter Zumthor in Cologne. However in this context, I rather would like to mention another architects, who used bricks in a new way in his time. His name is Eladio Dieste. One of the few architects who managed to blur the boundaries between tradition and innovation. The building I am referring to is the church "Iglesia de Atlántida Cristo Obrero y Nuestra Señora de Lourdes".

He designed the church in 1958, long before the applied use of computers in architecture. Eladio Dieste created double curved walls and ceilings by bricks, which are still today stunning and unique. This creates amazing light conditions and a friendly space, driven by the look and color of the bricks. Lastly I would like to refer to a book by the architect Koen Mulder "Het Zinderend Oppervlak", a book which highlights the new

A: 벽돌은 고대의 재료로 오늘날에도 재발명할 수 있다. 바로 떠오르는 프로젝트 중 하나는 피터 줌터(Peter Zumthor)가 독일 쾰른(Köln)에 지은 콜룸바 박물관(Kolumba Museum)이다. 그러나 이 맥락에서 또 다른 건축가를 언급하고 싶다. 그는 당시에 벽돌을 새로운 방식으로 사용했으며 그의 이름은 일라디오 디에스테(Eladio Dieste)이다. 전통과 혁신의 경계를 흐리게 만든 몇 안 되는 건축가 중 한 명이다. 내가 생각하고 있는 건물은 이글레시아 데 아틀란티다 크리스토 오브레로 이 누에스트라 세뇨라 데 로르데스(Iglesia de Atlántida Cristo Obrero y Nuestra Señora de Lourdes) 교회이다.

그는 건축에서 컴퓨터를 사용하기 훨씬 전인 1958년에 이 교회를 디자인했다. 일라디오 디에스테가 벽돌로 만든 이중 곡선 모양의 벽과 천장은 오늘날에도 여전히 놀랍고 독특하다. 이는 벽돌의 모양과 색상에 의해 구동되는 놀라운 빛과 우호적인 공간을 만든다.

possibilities in pattern and curvature for walls made out of bricks.

마지막으로 나는 건축가 코엔 멀더(Koen Mulder)의 "Het Zinderend Op'pervlak"이라는 책을 언급하고 싶다. 이 책은 벽돌로 만든 벽에 패턴과 곡선의 가능성을 강조한다.

Iglesia de Atlántida Cristo Obrero y Nuestra Señora de Lourdes
©Andrés Franchi Ugart

Kolumba Museum ©HOWI

BRICK-Q2: What are the strengths and weaknesses of brick?

A: **Bricks can be seen an nearly any culture.** They can be mass-produced, or customized. They give architecture a native look. They can with assistance by algorithms and robotics create wonderful patterns, if we refer to the works by Gramazio & Kohler, which have been possible since ages by bricklayers, but most of the techniques are forgotten. Nowadays, robots are applying these techniques faster and even with a high accurately.

A: 거의 모든 문화에서 벽돌을 찾아볼 수 있다. 대량 생산하거나 맞춤형으로 만들 수 있고, 건축에 태생의 모습을 준다. 그래마지오와 콜러(Gramazio & Koehler)의 작품에서 볼 수 있듯이 알고리즘과 로봇공학의 도움을 받아 훌륭한 패턴을 만들 수도 있다. 이는 벽돌공이 오랜 세월 동안 해온 일이지만, 기법 대부분이 잊혔다. 요즘은 로봇이 이러한 기술을 더 빠르고 정확하게 적용하고 있다.

Acoustic Bricks ©Gramazio Kohler Research, ETH Zurich

TILE-Q1: Tell us about your favourite project that you used tile in or another architect's work - interior, facade, etc.

A: Tiles are amazing in the past and today. They can have special surfaces that can create a unique look. Most people probably think about bathrooms or floors by mentioning tiles. But what if they are used on facades? They are durable and easy to clean. Architect Frank Lloyd Wright innovatively used three-dimensional tiles in such a way for the Enis house (1923), that the director Ridley Scott shot a scene of the famous futuristic film Bladerunner (1982) in that house. The entire system of the house is based on the tile pattern, which is located both inside and in the facade.

A: 타일은 과거에도 오늘날에도 놀라운 재료이다. 독특한 외관을 만들 수 있는 특별한 표면도 있다. 대부분의 사람은 타일이라고 하면 욕실이나 바닥을 생각할 것이다. 하지만 타일을 파사드에 사용한다면 어떨까? 타일은 내구성이 있고 청소하기 쉽다. 건축가 프랭크 로이드 라이트(Frank Lloyd Wright)는 에니스 하우스(1923년 - Enis House)에 3차원적인 타일을 굉장히 혁신적으로 사용했으며, 영화감독 리들리 스콧(Ridley Scott)이 여기서 유명한 초현대적인 영화 블레이드 러너(1982년)의 한 장면을 촬영할 정도였다.

Ennis House ©Mike Dillon

However, I would like to point out another architect: Manuel Herz. He used a special tile for the façade of the Jewish Community Center in Mainz, Germany. The building is clad with three-dimensionally riffled, green glazed ceramic tiles, which underlines the sculptural appearance of the center. A special assembly plan has been created which creates a pattern of different standardized lengths of tiles. Lastly, a pavilion, which very impressed me is the Austrian pavilion for the Expo by SPAN & Zeytinoglu Architects in 2010. The organic shape and the gradient by using white and red tiles was an amazing idea. It also shows, that organic shapes can be covered by tiles. In summary, can be applied to facades, ceilings etc. and can have different forms, which in the age of Industry 4.0 and mass-customization creates new possibilities for new forms.

집 전체의 시스템은 내부와 외관 모두에 있는 타일 패턴을 기반으로 한다.

하지만, 여기서 나는 또 다른 건축가인 마누엘 헤르츠(Manuel Hertz)를 언급하고 싶다. 그는 독일 마인츠(Mainz)에 있는 유대인 커뮤니티 센터의 파사드에 특별한 타일을 사용했다. 잔물결을 3차원적으로 만든 녹색 유약 세라믹 타일로 건물을 덮었으며, 이 타일은 센터의 조각 같은 형태를 강조한다. 이 프로젝트를 위해 특별 생산 계획을 세워 표준화된 여러 길이의 패턴을 만들 수 있었다. 마지막으로, 2010년 스판 앤 제이티노글루 건축가(SPAN & Zeytinoglu Architects)가 엑스포를 위해 디자인한 오스트리아 파빌리온에서도 깊은 인상을 받았다. 흰색과 빨간색 타일을 사용하여 만든 유기적인 모양과 그라디언트는 놀라운 아이디어였다. 또한 타일로 유기적인 모양을 덮을 수 있다는 점을 보여준다. 요약하자면, 타일은 파사드, 천장 등에 적용할 수 있으며 다양한 형태를 띨 수 있다. 4차 산업 및 대량 맞춤화의 시대에는 새로운 형태에 대한 새로운 가능성을 창출한다.

062 Brick, Brick! What do you want to be?

Austria Pavilion ©KimonBerlin

TILE-Q2: What are the strengths and weaknesses of tile?

A: As a building material, ceramic tiles are very ecologically friendly to the environment. They are really just clay, some minerals, water and fire. Ceramic tiles usually have a dense surface such as window glass (eg. porcelain stoneware) and are often very smooth. For this reason, dirt and germs on ceramic tiles have little chance. In general they are standardized rectangular patterns, which can be positive or negative.

A: 건축 자재로서 세라믹 타일은 매우 친환경적이다. 타일은 따지고 보면 그저 점토, 미네랄, 물, 그리고 불이다. 세라믹 타일은 일반적으로 유리창과 같은 고밀도 표면을 가지고 있고 (예: 도자기 석기) 종종 아주 매끄럽다. 이런 이유로 세라믹 타일에는 흙과 세균이 거의 없다. 일반적으로 타일은 포지티브이든 네거티브이든 직사각형 패턴으로 표준화되어있다.

GLASS-Q1: Tell us about your favourite project that you used glass in or another architect's work - interior, facade, etc.

A: **As a design element of contemporary architecture, the building material glass is almost indispensable.** The continuous development of manufacturing and processing technology and the constant improvement of ready-to-use products play a crucial role. Except for a special architecture, like military bunker, glass is used in all buildings.

A: **현대 건축의 디자인 요소로서 유리는 거의 필수 불가결하다.** 제조 및 가공 기술의 지속적인 개발과 기성 제품의 지속적인 개선이 중요한 역할을 한다. 군용 벙커와 같은 특수 건축물을 제외하고 유리는 모든 건물에 사용된다.

One famous building is the glass house by the American architect Philip Johnson, who wanted a house made entirely of glass and consisting of only one large room. In this way he wanted to break the boundaries between inside and outside. This characteristic leads me to a new architecture based on glass (fibers). The World Wide Web. I think most beings do not see this as a building, but it is a built structure that determines our way of life. Like the glasshouse by Johnson, it breaks the boundaries between inside and outside, which is in conclusion often then controlled by governments. But like a sewage system, which connects all buildings in a city, the WWW connects through Fibre-to-the-Home (FTTH) all buildings to a global network-city; hence it is architecture and we need to develop a new consciousness, of how we, as architects plan and design in such a city. A city, which is based on glass (and data).

유리로 된 유명한 건물 중 하나는 미국 건축가 필립 존슨(Philip Johnson)이 디자인한 유리 집이다. 그는 큰 방 하나로 구성되고 완전히 유리로만 만든 집을 원했다. 이 방식으로 그는 내부와 외부 사이의 경계를 깨고 싶어 했다. 이 특성은 나를 유리(섬유)를 기반으로 한 새로운 건축 양식으로 이끈다 - 월드 와이드 웹(World Wide Web). 대부분은 이를 건물로 보지 않지만, 나는 우리의 삶의 방식을 결정하는 구조라고 생각한다. 존슨의 유리 집과 마찬가지로 내부와 외부의 경계를 깨고 결론적으로 정부에 의해 통제된다. 그러나 도시의 모든 건물을 연결하는 하수 시스템과 마찬가지로 WWW는 모든 건물을 가정 광가입자망(Fiber to the Home, FTTH)을 통해 글로벌 네트워크의 도시로 연결한다. 그러므로 WWW는 건축이며 우리는 유리(와 데이터)에 기반한 이런 도시에서 계획하고 디자인하는 건축가로서 새로운 의식을 키울 필요가 있다.

GLASS-Q2: What are the strengths and weaknesses of glass?

A: The raw materials for glass are practically unlimited available. However, it is relatively sensitive to

A: 유리의 원료는 사실상 무제한이다. 하지만 구체적인 목적으로 만들지 않으면 기계

mechanical stress, if not specific made. Probably a downside of glass is its cost, especially when it comes to special glass constructions.

적 스트레스에 상대적으로 민감하다. 유리의 단점은 비용일 것이며 특히 특수 유리 구조일 때 더 그렇다.

WOOD-Q1: Tell us about your favourite project that you used wood in or another architect's work - interior, facade, etc.

A: Well, since this is a book about materials and probably they are countless buildings on the planet, many of them made of wood, I would like to compare two projects, which are from different times, but the same technique.

A: 이 책은 재료에 관한 책이고, 아마 지구상에는 수많은 건물이 있으며, 그중 많은 수가 나무로 만들어졌을 것이기 때문에 나는 같은 기술을 쓰지만 다른 시대의 프로젝트 두 개를 비교하고 싶다.

Glass House ©Staib

066 Brick, Brick! What do you want to be?

On the one hand, there is the Naiju Residential Center and Kindergarten in Chikuho, Fukuoka, by Shoei Yoh(1995), who uses a special grass: bamboo.

On the other side is the Center Pompidou in Metz(2010) by Shigeru Ban and Jean de Gastines with a timber-beam structure. Both projects are a bamboo / wood matrix that creates a freeform that is clad with another material. The differ in the use of computers. In 1995, there weren't really computers or software's available, which could compute such a complexity. I am not sure, but I guess that they used models, like Frei Otto (Multihalle Mannheim, 1975) in its time. Shigeru Ban had the possibility to collaborate with the Swiss/German company Designtoprodtuction, which generated an algorithm, which directly communicated with the machines. With this algorithm it was possible to CNC mill 18.000 meters of the timber-structure and to produce about 1,800 double-curved glulam segments. So we see that certain experimental wood constructions have been possible for decades, but it becomes feasible through the use of algorithms, which reflect the idea of mass-customization. Also we have to admit, that bamboo and wood are

하나는 요우 쇼헤이(1995년)가 후쿠오카 치쿠호에 지은 나이주 주거 센터 및 유치원이다. 그는 이 건물에 특별한 식물을 썼는데, 이는 바로 대나무이다.

다른 하나는 반 시게루(Shigeru Ban)와 장 드 가스틴(Jean de Gastines)이 메츠(2010년)에 목재 들보 구조로 디자인한 폼피두 센터이다. 두 프로젝트 모두 다른 재료로 덮고 자유형을 만드는 대나무/나무 매트릭스이다. 차이가 있다면 컴퓨터이다. 1995년에는 그런 복잡성을 계산할 수 있는 컴퓨터나 소프트웨어가 실제로 존재하지 않았다. 확실하지는 않지만, 당시에 프레이 오토(Frei Otto - Multihalle Mannheim, 1975)와 같은 모델을 사용했던 것 같다. 반 시게루는 기계와 직접 통신하는 알고리즘을 생성한 스위스/독일 회사인 Designtoproduction과 협력할 수 있었다. 이 알고리즘을 사용하여 CNC 밀로 18m의 목재를 다듬고 약 1,800개의 이중곡선 글루렘 부분을 생산할 수 있었다. 그래서 실험적인 목재 구조가 수십년 동안 가능했음을 알 수 있지만, 알고리즘을 사용하여 실현 가능성이 커졌으며 이는 대량 맞춤화를 반영한다. 또한 대나무와 나무가 서로 다른 정체성을 가진 두 가지 재

Center Pompidou in Metz ©Dietmar Köring

two materials with different identities. Bamboo itself has a history in Asia, but not really in Europe or America, but it should be rethought, as it is a perfect building material. Perhaps it would also be ideal to renature opencast mining areas in Europe. It could be a unique business model that replaces coal with fast growing green bamboo.

WOOD-Q2: What are the strengths and weaknesses of wood?

A: Wood is replenished daily by nature in large quantities. With sustainable forestry and responsible consumption, sufficient wood is available for all building purposes at all times. Wood, when properly processed and used purposefully, offers a durability that far exceeds that of most other products. Some wooden buildings of the last 200 to 500 years are often still standing. However, wood is flammable and can be attacked by pests, bacteria or insects. Bamboo impressed above all by its enormous growth speed. Bamboo is resistant and durable even under heavy use. It is very light and elastic thanks to the cavities in the cell structure.

료라는 것을 인정한다. 대나무는 아시아에서 역사가 있다. 실제로 유럽이나 미국에서는 그다지 쓰이지 않지만 완벽한 건축 자재이기 때문에 그 가능성을 재고해봐야 한다. 어쩌면 유럽의 오픈 캐스트 광산 지역을 대나무로 복원하는 것이 이상적일지도 모르겠다. 빠르게 성장하는 친환경적인 대나무로 석탄을 대체하는 독특한 비즈니스 모델이 될 수도 있다.

A: 나무는 자연이 매일 대량으로 보충한다. 지속가능한 임업 및 책임있는 소비로 모든 건설 목적을 위한 목재는 항상 충분할 수 있다. 목재는 적절하게 가공하고 의도적으로 사용한다면 다른 재료 대부분의 내구성을 훨씬 뛰어 넘는다. 지난 200년에서 500년 사이에 지어진 목조 건물 일부는 종종 여전히 서 있다. 그러나 나무는 가연성이 높으며 해충, 박테리아 또는 곤충에 의해 공격받을 수 있다. 대나무는 무엇보다도 그 엄청난 성장 속도가 아주 인상적이다. 대나무는 험한 상황에서도 내구성이 강하고 튼튼하다. 또한 세포 구조의 구멍 덕분에 매우 가볍고 탄력적이다.

AZC

Who is...?

AZC was founded in 2001 with the idea that exploring architecture and its techniques could help to improve our built environments. Our interest does not lie in inventing concepts, we have always sought to realize buildings for real life's needs.

Through competitions and direct commissions, our office has worked on over a hundred projects of varied scales and uses. Most of our built projects are intended for a wide audience; sports facilities, lecture halls, office buildings and residential, some of which very specific for vulnerable populations.

Q1: What is material to an architect
     (or to you)?

A: **To us, material is the language.** It is what we assemble to make our buildings come to life.

A: 우리에게 재료란 언어이다.
우리가 건물에 생명을 불어넣기 위해 조립하는 것이 재료이다.

Q2: Tell us about your favourite
     (or most often used) material and why.

A: We use all kinds of materials, from concrete to PVC. We like brick, glass and wood.

A: 우리는 콘크리트부터 PVC까지 모든 종류의 재료를 쓰고 벽돌, 유리와 나무를 좋아한다.

We use all kinds of materials

Q3: When do you decide the material during the design process and what is your criteria? (e.g. budget, client's preference, design concept, climate, etc.)

A: When we start working on a project we think of the materials too.

A: 우리는 프로젝트를 시작할 때 재료도 생각한다.

Q4: What are some architectural projects that inspired you regarding brick, tile, wood and/or glass? And why?

A: Binet: The daylight factories.
"In 19th Century America, industrial buildings were built similarly to their residential counterparts: a post and beam structural system in brick and timber generally enveloped by brick or stone cladding.
The industrial appearance of these buildings came from their height and lack of ornamentation rather than from any real structural distinction. However, this structural system eventually limited the size of the industrial buildings so that in many respects industrial and residential architecture from that period hardly differed.
Expanding the applications of reinforced concrete after 1903, american architect

A: 비네 오피스 지구 (Hôtel d'entreprises Binet): 일광 공장
19세기 미국에서 산업용 건물은 주거용 건물과 유사하게 지어졌다. 벽돌과 목재로 된 기둥-보 구조를 일반적으로 벽돌이나 돌 클래딩으로 둘러쌌다. 이런 건물의 산업적 외관은 실제 구조적인 차이보다는 그 높이와 적은 장식에서 비롯되었다. 그러나 이 구조는 결국 산업 건물의 크기를 제한하여 많은 면에서 그 시대의 산업 및 주거용 건축물은 서로 거의 다르지 않았다. 1903년 이후 철근 콘크리트의 적용을 확대한 미국 건축가 앨버트 칸(Albert Kahn)은 디트로이트의 이스트 그랜드 대로 (East Grand Boulevard)에 패커드 자동차(Packard Motor

01 - BETON
02 - LAME D'AIR
03 - ISOLANT THERMIQUE
     LAINE MINERALE
04 - CONTRE VOILE - BRIQUE PLEINE
05 - BSO (BRISE SOLEIL ORIENTABLE)
06 - ENCADREMENT BAIE
     ALUMINIUM ANODISE
07 - GARDE-CORPS
08 - MENUISERIE ALUMINIUM
     ANODISE
09 - PARE PLUIE

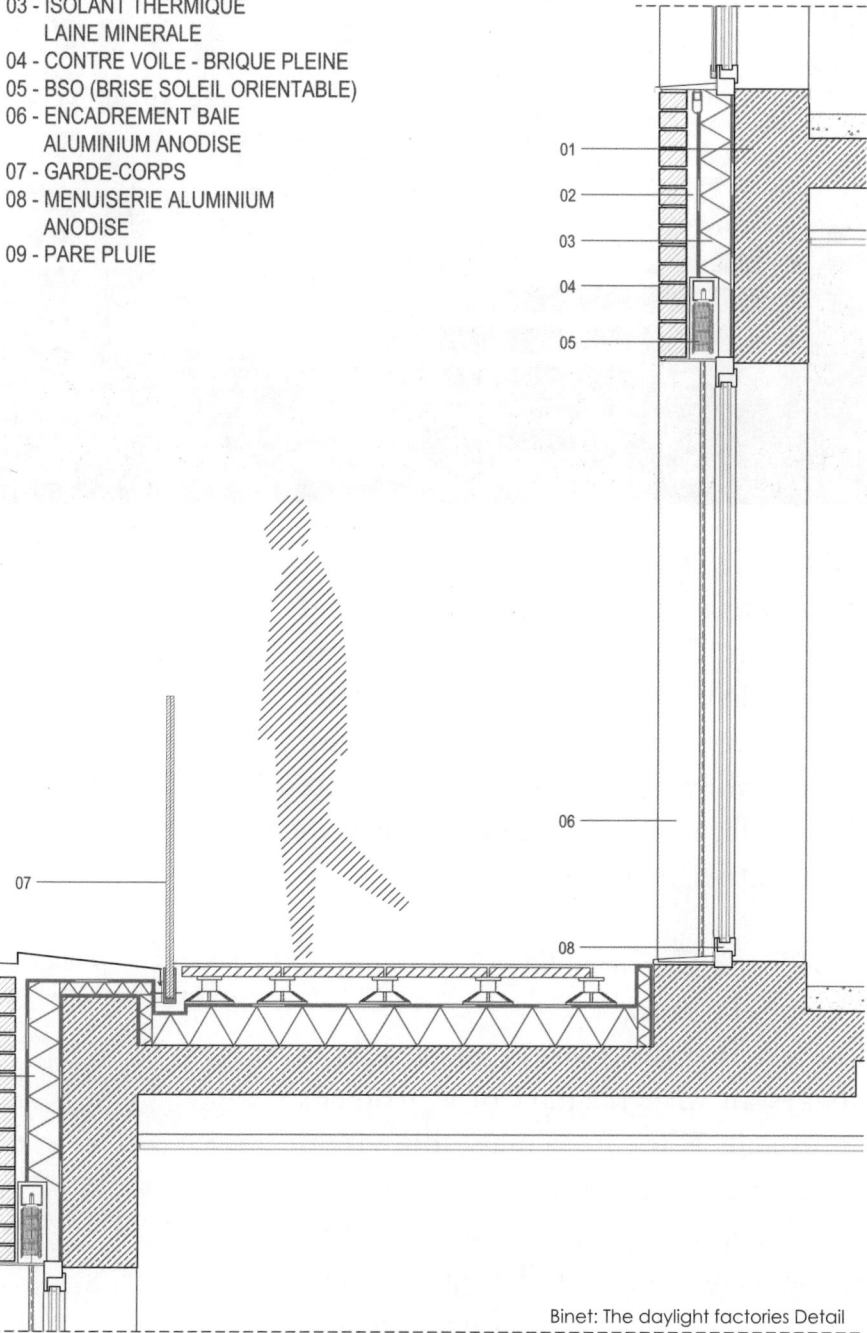

Binet: The daylight factories Detail

Binet: The daylight factories.

Albert Kahn built the Packard Motor Car Company on East Grand Boulevard in Detroit. There the reinforced concrete frame held the loads so that the perimeter walls of the factory could be filled with glass to allow natural light to penetrate the interior workspaces, thus giving birth to the "Daylight Factory."

NDBS and Pacemar: The O Museum in Myajima by Sanaa.
For us, thinking about materials is always an integral part of a project's approach. With its remarkable visual qualities, of reflection, transparency and disappearance, glass adds versatility, strength and durability. The glass facades stand out for their ease of maintenance

Car Company)를 건설했다. 철근 콘크리트 프레임으로 하중을 유지하고, 채광이 내부 작업 공간에 들어오도록 공장의 주변 벽을 유리로 채워 "일광 공장(Daylight Factory)"이 탄생했다.

NDBS와 페이스마(Pacemar) SANAA의 오가사와라 박물관 (O Museum)
우리에게 재료에 대해 생각하는 것은 프로젝트를 디자인할 때 항상 필수적인 부분이다. 반사, 투명성 및 소멸 같은 놀라운 시각적 특성으로 유리는 다용성, 강도 및 내구성을 더한다. 유리 파사드는 유지 보

and are insensitive to the weather: rain, snow or ice have no grip on them. Cleaning is easy, maintenance is reduced; and at the end of the life cycle, the glass is recyclable.

Neudorf: The wooden barns

Wood is environment friendly, renewable and consumes the least amount of energy when processed. Also, very important for the wellbeing of the occupants of the buildings, wood has a pleasantly fresh smell and is also capable of absorbing unpleasant odors: because wood breaths, it acts as a filter that cleans and moistens the air. As a building material wood allows industrial fabrication of big pieces and the rapid installation on site.

수의 용이성이 뛰어나며 날씨에 민감하지 않다. 유리는 비, 눈 또는 얼음에 크게 영향 받지 않는다. 청소가 쉽고, 유지 보수가 적으며, 다 쓴 후 재활용할 수 있다.

김나지움(Sports center in Neudorf) : 나무 외양간
목재는 환경친화적이고 재생 가능하며 가공 시 가장 적은 양의 에너지를 소비한다. 또한, 건물 거주자의 복지에 매우 중요한 점인데, 목재는 쾌적하고 신선한 향기가 나고 불쾌한 냄새를 흡수할 수 있다. 나무는 호흡을 하므로 공기를 깨끗하고 촉촉하게 하는 필터 역할을 한다. 건축 자재로서 목재는 큰 조각을 공업적으로 제작할 수 있고 현장에서 신속한 설치가 가능하다.

Terrasse 9: Lisbon buildings

NDBS and Pacemark The O Museum in Myajima by Sanaa.

Terrasse 9: Lisbon buildings

The tiles used on Terrasse 9, are hand made by a small factory called "Rairies Montrieux", one of the latest "Artisan Briquetier" to keep precious local and ancestral techniques of shaping tiles and bricks terracotta old. Holder of an authentic workshop, where nothing has changed for decades, the "Rairies Montrieux" work with the heritage, memory of know-how and authenticity for quality tiles.

테라스 9(Terrasse 9): 리스본의 건물들
테라스 9에 사용된 타일은 "레이리 몽리유(Rairies Montrieux)"라는 작은 공장에서 직접 제작한 것이다. 이 공장은 귀중한 현지와 선조의 타일 및 벽돌 테라코타 기술을 그대로 유지하는 최신 "벽돌 장인(Artisan Briquetier)" 중 하나이다. 수십 년 동안 아무 것도 변하지 않은 진정한 작업장의 소지자인 레이리 몽리유는 전통, 노하우의 기억 및 품질 높은 타일의 진위성을 이어간다.

Q5: Tell us about the materials you are interested in or want to use in your projects right now.

A: We would like to use more plain brick, as a wall. And wood as walls and structure.
As we try to work with the same artisans on several projects, we would like to allow for more innovation and experimentation.

A: 우리는 평범한 벽돌을 벽에 더 써보고 싶다. 그리고 나무도 벽이나 구조로 써보고 싶다. 우리가 몇몇 프로젝트에서 같은 장인과 함께 일하려고 할 때, 우리는 더 많은 것들을 실험해 보고싶다.

BRICK-Q1: Tell us about your favorite project that you used brick in or another architect's work - interior, facade, etc.

A: We used brick on the last project of an office building: "Hotel industriel Binet" in Paris.
The white glazed brick façade fits with the north area of Paris where the building belongs and with its function of an office building.

A: 파리의 '비네 오피스 지구 (Hôtel Indsutriel Binet)'에 있는 마지막 사무용 건물에 벽돌을 썼다. 하얀 유약처리가 된 벽돌 파사드는 건물이 있는 파리의 북부 지역과 사무용 건물이라는 용도에 잘 맞았다.

BRICK-Q2: What are the strengths and weaknesses of brick?

A: **Brick is a very graphic material as the light showcases every brick and draws every joint.**
It also is a long-lasting material with no need for refurbish or clean.
It would be nice if we could use it as a structural material for columns or beams also.
It would be interesting to use it as an insulating material also.

A: 벽돌은 빛이 비추면 모든 벽돌이 잘 보이고 모든 이음매가 눈에 띄기 때문에 매우 그래픽한 재료이다. 또한 보수나 청소없이도 오래가는 재료이다. 기둥이나 들보 같은 구조에도 사용할 수 있다면 좋을 것 같고 절연재로 사용하는 것도 흥미로울 것 같다.

TILE-Q1: Tell us about your favourite project that you used tile in or another architect's work - interior, facade, etc.

A: We used glazed tiles on the exterior cladding of "Terrasse 9" housing project.

A: 우리는 '테라스 9 (Terrasse 9)'라는 주택 프로젝트의 외장 마감재로 채유 타일(표면에 유약을 입힌 타일)을 썼다.

TILE-Q2: What are the strengths and weaknesses of tile?

A: Tile is a material that does not need maintenance. It lasts for a long time in excellent condition.

A: 타일은 특별한 관리가 필요 없는 재료이다. 훌륭한 상태로 오랫동안 지속된다.

Terrasse 9

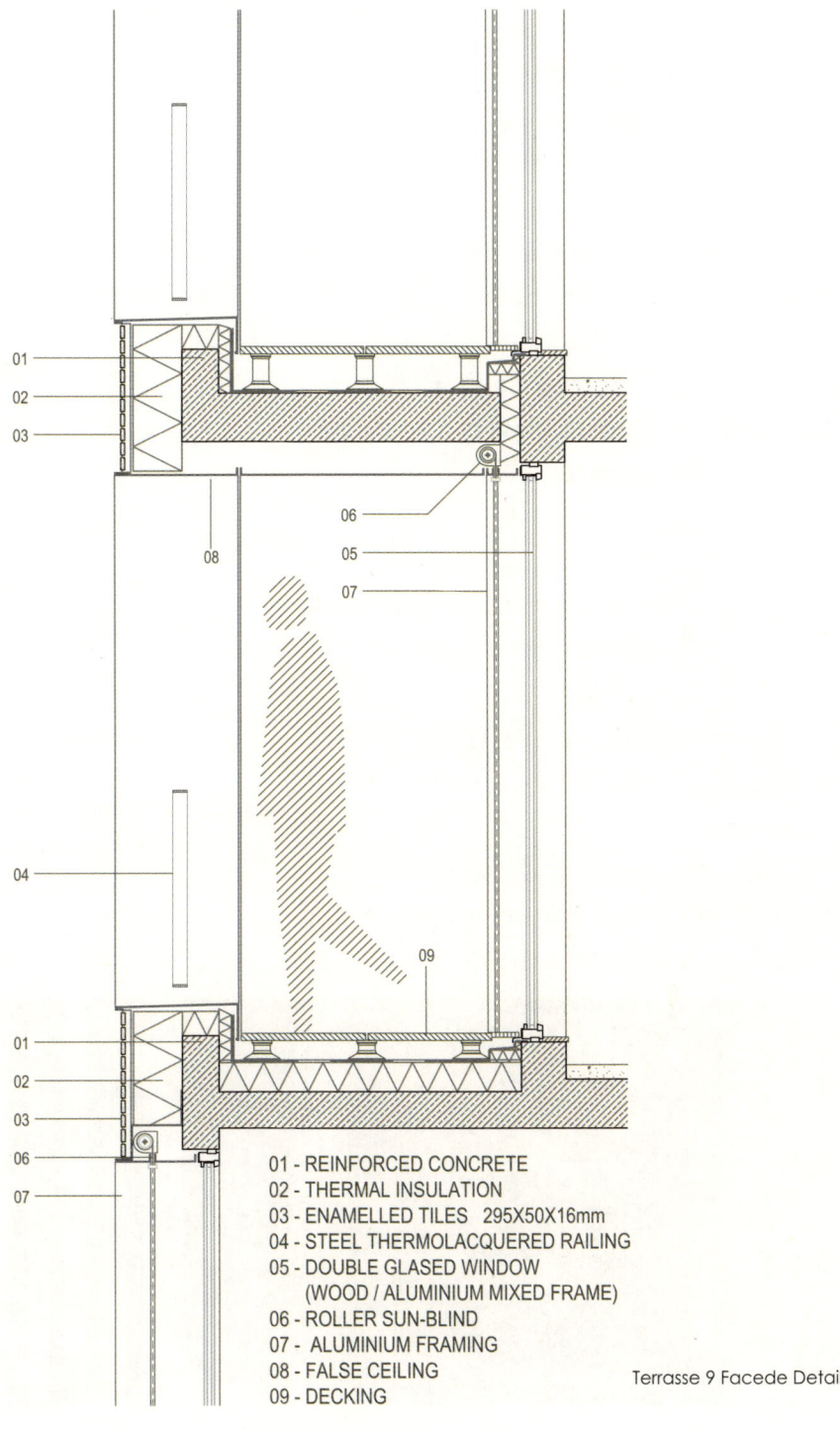

GLASS-Q1: Tell us about your favourite project that you used glass in or another architect's work - interior, facade, etc.

A: We used glass on a medical house project "NDBS" in Paris and on an office building "Pacemar" in Suresnes. For NDBS project, glass is a cladding material. For Pacemar, glass is a ventilated double façade material.

A: 우리는 파리에 있는 'NDBS(Notre-Dame de Bon Secours)'라는 의료시설과 프랑스 쉬렌에 있는 페이스마(Pacemar)의 사무건물 (Louis Dreyfus Armateurs Headquarters)에 유리를 썼다. NDBS 프로젝트에서 유리는 마감재이다. 페이스마의 건물에서 유리는 환기가 되는 이중 파사드 재료이다.

GLASS-Q2: What are the strengths and weaknesses of glass?

A: The strengths are the transparence, the lightness, the long-lasting with no maintenance.
Weakness, the price and the fact that in some extreme temperature areas, hot and cold it does not protect the interior.

A: 장점은 투명도, 밝기와 관리 없이 오래가는 점이다. 단점은 가격과 뜨겁거나 추운 일부 극한 온도 지역에서 내부를 보호하지 못한다는 점이다.

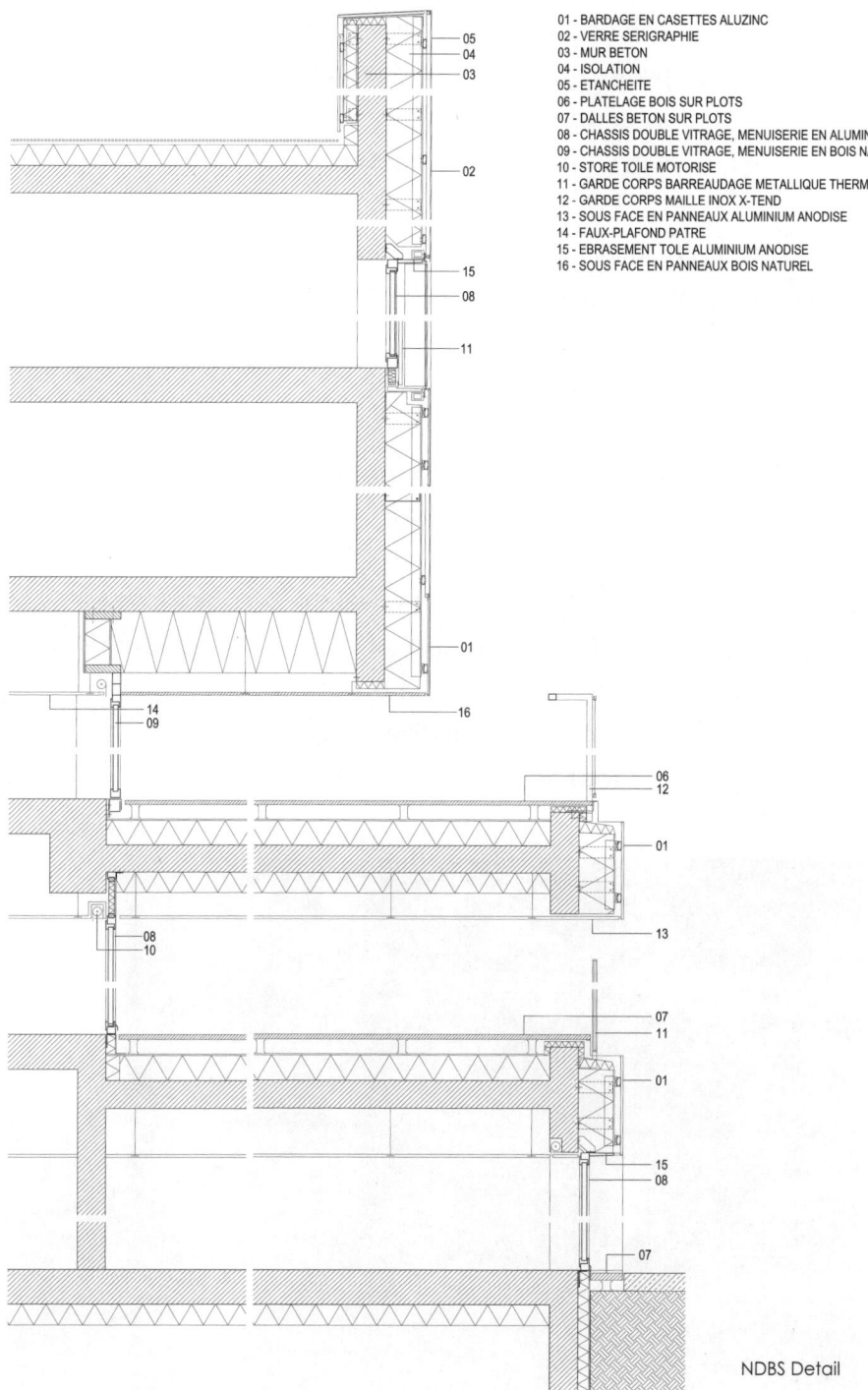

WOOD-Q1: Tell us about your favourite project that you used wood in or another architect's work - interior, facade, etc.

A: The "Gymnase" is a sports hall in Strasbourg. We massively used wood in this project.

A: '짐나지움 (Gymnase - Sports centre in Strasbourg)'은 프랑스 스트라스부르에 있는 체육관이다. 우리는 이 프로젝트에 나무를 대량으로 썼다.

WOOD-Q2: What are the strengths and weaknesses of wood?

A: **Is an ecological material, very warm and human, with a low CO2 impact.** It does need protection when used on the exterior. It does need fire protection when used as a main structure.

A: 나무는 이산화탄소가 낮은 환경 친화적인 재료이고, 매우 따뜻하며 인간적이다. 하지만 외관에 사용할 때는 보호 처리가 필요하고 주요 구조로 사용할 때는 화재 방지 처리가 필요하다.

Gymnase-Sports centre in Strasbourg

Gymnase-Sports centre in Strasbourg

BOARD

Who is ...?

BOARD (Bureau of Architecture, Research, and Design) was founded in Rotterdam in 2005 and is active in many fi elds: as an architecture, urban design, and design practice, as a research board and as a platform for comparative analysis on urban issues through its bi-annual journal MONU – Magazine on Urbanism. BOARD won several prizes recently in prestigious international architecture and urban design competitions.

Q1: What is material to an architect
(or to you)?

A: **To me material is usually the final physical manifestation and embodiment of an idea or a concept for a project.** In most cases, and especially in architectural competitions, it is the last thing that has to be decided on. What I am interested in is that the material of a project is as tightly as possible related to the original idea of a project, and translates this original idea into the space as literally as possible. Only then, I believe, a material acquires a bigger meaning and develops the power to create objects, spaces, and shelters that support and improve our lives, which is at the end of the day the most basic function of architecture and design.

A: 나에게 재료는 일반적으로 아이디어의 최종적인 물리적 표현과 구현 또는 프로젝트의 콘셉트이다. 대부분의 경우, 특히 건축 공모전에서는 마지막에 결정하는 사항이다. 내 관심사는 프로젝트의 재료가 원래 아이디어와 최대한 가깝고, 이 아이디어를 가능한 한 문자 그대로 공간으로 번역하는 것이다. 나는 그럴 때만 재료가 더 큰 의미를 지니고, 우리의 삶을 지원하고 개선하는 물체, 공간과 피난처를 만드는 힘을 얻는다고 믿는다. 이는 결국 건축과 디자인의 가장 기본적인 기능이다.

Q2: Tell us about your favourite
(or most often used) material and why.

A: Although I don't have a particular favourite material - as it depends always on the project and its relevance in relation to the conceptual idea of project - I am quite a fan of styrofoam and in particular blue styrofoam that comes usually in the form of boards. What fascinates me

A: 재료는 프로젝트 자체와 아이디어와의 적절함에 달려있기 때문에 특별히 좋아하는 재료는 없지만, 나는 스티로폼, 특히 보드 형태로 나오는 파란색 스티로폼의 팬이다. 이 재료의 매력적인 부분은 건축의 가장 흥미진진한 부분 중 하나인 창조 과정을 보여준다는 점

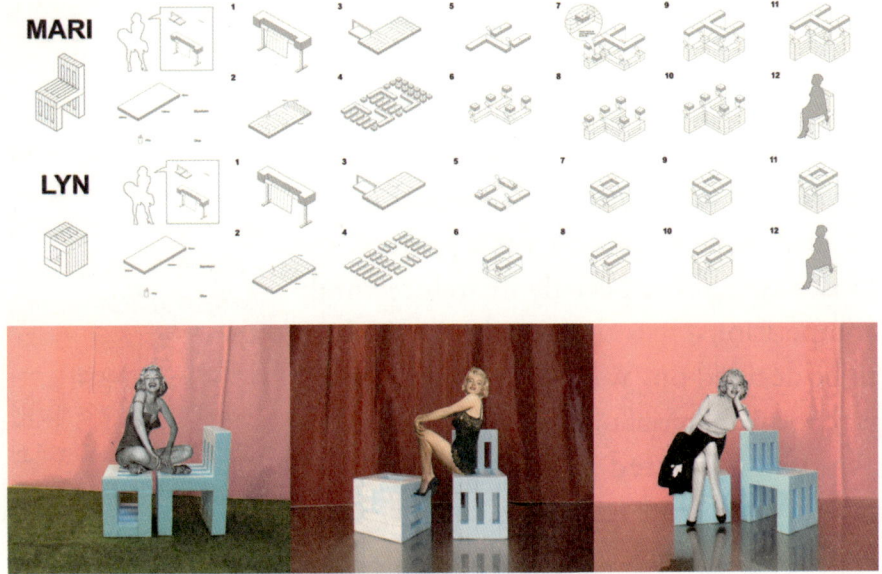

Mari and Lyn

about this material is that it represents to me one of the most exciting parts of architecture: the process of creation. Because every new project we usually start with an enormous amount of testing with scale models. And this trial and error design method is always processed in blue styrofoam that we purchase at a local hardware store here in Rotterdam and shape with a hot-wire foam cutter. **Our project Mari and Lyn - two very simple and very affordable pieces of furniture, a chair and a stool, that can be self-built** - has to be seen as an homage to this beloved and extraordinary material.

이다. 새로운 프로젝트마다 우리는 대개 스케일 모형으로 엄청난 양의 테스트를 시작한다. 그리고 이 시행착오는 항상 로테르담의 한 동네 철물점에서 사는 파란색 스티로폼으로 가공하고 핫와이어로 모양을 만든다. **우리 프로젝트인 마리와 린(Mari and Lyn)은 매우 간단 및 저렴하고 혼자 조립할 수 있는 가구 두 개로, 의자와 스툴로 구성되어 있다.** 이는 이 사랑스럽고 놀라운 재료에 대한 오마주로 볼 수 있다.

Q3: When do you decide the material during the design process and what is your criteria? (e.g. budget, client's preference, design concept, climate, etc.)

A: The design project Mari and Lyn shows that it happens sometimes that the choice for a particular material is made even before the designing starts. However, in our work this is an exception. But in the case of Mari and Lyn the material came first and influenced the design with its characteristics, size, and availability. Because with the project we wanted to celebrate blue styrofoam not only as a great model-making material that is very light – styrofoam consists of ninety-eight percent air – highly stiff and very easy to cut, that swims while resisting moisture, and is relatively cheap, but also as a great architectural design-tool in general that became world-famous after having been used by the Dutch architectural avant-garde of the 1980s and 1990s.

A: 디자인 프로젝트 마리와 린은 디자인을 시작하기 전에 특정 재료를 선택하는 경우가 종종 있음을 보여준다. 하지만 우리 일에서 이것은 예외이다. 마리와 린의 경우 재료를 먼저 결정하여 디자인의 특성, 크기 및 가용성에 영향을 미쳤다. 이 프로젝트를 통해 우리는 파란 스티로폼이 매우 가볍고 – 스티로폼은 98%가 공기이다 – 매우 단단하고 매우 자르기 쉬우며, 내습성이 있지만 수영을 할 수 있고, 비교적으로 저렴한 훌륭한 모델 만들기용 재료일 뿐만 아니라 훌륭한 건축 디자인 도구임을 찬양하고 싶었다. 이는 1980년대와 1990년대에 네덜란드 건축의 아방가르드 건축가들이 쓴 이후로 세계적으로 유명해졌다.

Q4: What are some architectural projects that inspired you regarding brick, tile, wood and/or glass? And why?

A: To a certain extend the tiles that we

A: 우리가 '사무실과 호텔 방'

BOARD 091

used to clad and camouflage the entrance door to the "Office and Hotel Room" were inspired by the famous tiles and grids of Superstudio. When we entered the building of the project the first time, we were intrigued by the existing tiles of the staircase - which many people of the building considered ugly and thus wanted to get rid of, but which we liked particularly because of their reminiscence to projects of Superstudio. Therefore, we wanted to do something with them right from the start. The negative attitude of the neighbours towards the tiles also triggered our interest in them. What we always liked about Superstudio is how they used the tools of art and literature, but also the rhetorical devices of metaphor and allegory, and the tools of irony and imagination in their projects, "manoeuvring through the no-man's-land between art and architecture so as to attempt forays into politics, sociology and philosophy", as Adolfo Natalini, one of the original founders of Superstudio" once described it in an interview called "Deadly Serious" that was published in issue #14 of our publication "MONU – Magazine on Urbanism".

입구 문을 덮고 위장하는데 사용한 타일은 슈퍼스튜디오의 유명한 타일과 격자에서 어느 정도 영감을 얻었다. 처음 그 건물에 들어갔을때 우리는 계단의 기존 타일에 흥미를 느꼈다. 건물의 많은 사람이 이 타일을 추악하다고 생각하여 철거하기를 원했지만, 우리는 슈퍼스튜디오 프로젝트를 회상하게 하는 점이 특히 마음에 들었다. 그래서 처음부터 이 타일로 뭔가를 하고 싶었다. 타일에 대한 이웃 사람들의 부정적인 태도 또한 우리 관심을 끌었다. 우리가 슈퍼스튜디오에 대해 항상 좋아했던 점 중 하나는 예술과 문학을 사용하는 방법뿐만 아니라 은유의 우화의 수사학적인 장치, 그리고 아이러니와 상상력을 사용하는 방법이었다. 슈퍼스튜디오의 원래 설립자 중 하나인 아돌포 나탈리니(Adolfo Natalini)는 이를 "정치, 사회학 및 철학에 대한 진출을 위해 예술과 건축 사이의 황무지 사이를 조심히 움직이는" 것이라고 우리 퍼블리케이션 "MONI – 어바니즘 매거진"의 이슈 14번에 출간된 "너무 진지함"이라는 인터뷰에서 옛날 말했었나.

Lately, we have become fascinated by ice as a building material.

최근 우리는 건축 자재로서 얼음에 매료되었다.

Q5: Tell us about the materials you are interested in or want to use in your projects right now.

A: **Lately, we have become fascinated by ice as a building material.** What we like about it is on the one hand its semi-transparent character that, if used as a wall, allows you to look through it, but not entirely, which adds a kind of mysteriousness to the spaces behind, an effect that is similar to translucent glass. But what we like most about it, is its instability when comes to its form when exposed to temperatures above zero degrees Celsius.
This appealing ephemeral character and aspect of ice we used when we recently designed a floating theatre on the river Spree in Berlin that was created out of blocks of ice. And since this stage only melts after a couple of days in the water, every performance becomes an ephemeral event unfolding into a unique moment in time that you don't want to miss. Accordingly, we called the project "A Unique Moment".

A: **최근 우리는 건축 자재로서 얼음에 매료되었다.** 우리가 좋아하는 것은 반투명한 성질인데, 벽으로 사용했을 때 들여다볼 수는 있지만, 완전히 다 볼 수는 없고, 벽 뒤의 공간에 일종의 신비감을 더한다. 반투명 유리와 비슷한 효과이다. 하지만 우리가 가장 좋아하는 것은 0도 이상의 온도에 노출되면 그 형태가 불안정하다는 점이다. 우리가 최근에 설계한 베를린의 슈프레 강에 떠다니는 극장을 얼음으로 만들었고 이때 얼음의 이 매력적인 일시적인 특성과 측면을 사용했다. 그리고 이 스테이지는 며칠만 지나면 물속에서 녹기 때문에 모든 공연은 놓치고 싶지 않은 특별한 순간이 펼쳐지는 일시적인 이벤트로 만들었다. 따라서 우리는 이 프로젝트를 '독특한 순간'이라고 불렀다.

Ice as a building material

BOARD 095

BRICK-Q1: Tell us about your favorite project that you used brick in or another architect's work - interior, facade, etc.

A: Until today, we have done only one project in which we used brick, which is the design of a square with urban furniture, a project that we called "Home Outside". There, we proposed to use brick for the surface of the square to connect it to its context. In that way the typical red brick surface of the neighbourhood could be continued and strong ties to the surroundings created. However, the main element of the square was not the surface, but 3 elements that we added, that are both artistic and functional: a fireplace, a table, and a boat. With these 3 elements we aimed to create a unique square as a work of art, where the inhabitants of the site, called the "Kraanbolwerk", but also its visitors, can feel at home outside and meet, talk, cook, eat, warm up, and play.

A: 오늘까지 우리는 우리 프로젝트에 단 한 번 벽돌을 썼었다. 이는 도시 가구가 있는 광장의 디자인으로, 우리가 "바깥의 집"이라고 부르는 프로젝트이다. 여기서 우리는 광장의 표면에 벽돌을 써서 그 콘텍스트에 연결할 것을 제안했다. 그렇게 하면 그 동네의 전형적인 붉은 벽돌 표면이 계속되면서 광장과 주변 환경 사이에 강한 유대 관계를 형성할 수 있었다. 그러나 광장의 주요 요소는 표면이 아니라 우리가 추가한 예술적이면서 기능적인 3가지 요소였다. 이는 벽난로, 테이블 및 보트였으며 이 세 가지 요소로 우리는 하나의 예술 작품인 독특한 광장을 만드는 것을 목표로 삼았다. "크란볼베르크(Kraanbolwerk)"라고 불리는 사이트의 주민들뿐만 아니라 방문객까지 바깥에서 집처럼 편안히 느끼면서 만나고, 얘기하고, 요리하고, 먹고, 몸을 따뜻이 하고, 놀 수 있게 하고 싶었다.

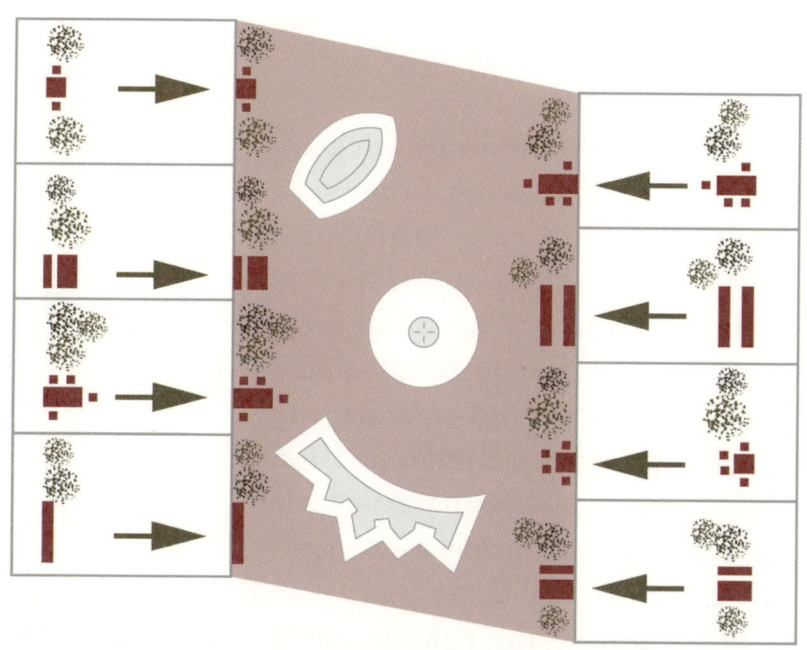

Home Outside

BOARD 097

BRICK-Q2: What are the strengths and weaknesses of brick?

A: Some of the strengths of brick are certainly its robustness, longevity, affordability, and ease of use and application. For the project "Home Outside" we wanted to use all of these qualities. But since the square was located on a former industrial site - a former city port, trading, and production location – where a renowned factory called "de Volharding", that was mainly constructed with brick, once stood, we wanted to connect the project through the use of brick for the square's surface with the industrial past of the place. In order to deal with one of the weaknesses of brick - namely its lack of reinforcement, making brick easy to assemble but also easy to disassemble – we placed the 3 elements of the square on concrete slabs with engraved in them information about the meaning behind the elements.

A: 벽돌의 장점 중 일부는 확실히 견고함, 수명, 경제성, 사용 및 적용 용이성이다. 프로젝트 '바깥의 집'을 위해 우리는 이 모든 특성을 사용하고 싶었다. 하지만 그 외에도 그 광장은 이전에 산업 현장으로 – 전 도시 항구, 무역 및 생산 장소였다 – 주로 벽돌로 지어진 "드 볼하딩(de Volharding)"이라는 유명한 공장이 있었기 때문에 벽돌을 사용하여 프로젝트를 사이트의 산업 과거와 연결하고 싶었다. 벽돌의 단점 중 하나는 보강물이 부족한 점이며, 이 덕분에 벽돌은 조립이 쉽지만 분해하기도 쉽다. 우리는 광장의 세 가지 요소를 콘크리트 슬래브에 배치하고 각 요소가 의미하는 바에 대한 안내정보를 새겼다.

TILE-Q1: Tell us about your favourite project that you used tile in or another architect's work - interior, facade, etc.

A: One of my favourite projects in which we used tiles is our design for an "Office and Hotel Room" that was recently completed in the centre of Rotterdam. There, since the entrance of the project was already slightly set apart from the main staircase towards a little niche, we saw an opportunity to create a "secret" and hidden door to the office and hotel room, which would disconnect the space magically and as a hideaway from the business of the city and 'real life' in general. To emphasize the hiddenness of the place, we continued the existing white ceramic tiles from the corridor into the niche of the entrance and covered the entire door with the same tiles too, which camouflages the entrance to such an extent that until today the postman is having trouble finding it.

A: 타일을 사용한 내가 가장 좋아하는 프로젝트 중 하나는 로테르담 중심부에서 최근에 완성된 "사무실과 호텔 방 (Office and Hotel Room)" 디자인이다. 프로젝트의 입구가 이미 작은 벽감을 향해 메인 계단과 약간 떨어져 있었고 우리는 이를 사무실과 호텔 방으로 향하는 "비밀"스럽고 숨겨진 문을 만들 기회를 보았다. 덕분에 마술처럼 공간을 분리하고 사무실과 호텔 방을 도시의 일상과 일반적인 '진짜 인생'으로부터 은신처로 만들 수 있었다. 이 공간을 더 잘 숨기기 위해 복도에 있는 기존의 흰색 세라믹 타일을 입구의 벽감까지 연결하고 문 전체도 같은 타일로 덮어서 입구를 위장했다. 오늘날까지도 우체부가 찾기 어려워할 정도다.

TILE-Q2: What are the strengths and weaknesses of tile?

A: Apart from being a hard-wearing material with an ability to cover surfaces in a simple, solid, and affordable way, we consider the ability of tiles to create complex mosaics - and with it narratives and multifaceted meanings - as one of its greatest strengths. This quality we wanted to use when we designed one of the lavatories for the "Office and Hotel Room" project. Because, while the main space of the project is white, open, neutral, and flexible, this lavatory contains an impressive hand-made piece of art depicting a palm tree, a 1:8 reproduction of a design originally made by STAR strategies + architecture and intended for a public square in the city of Elche, Spain. Thus, the clients, admirers of Adolf Loos, preferred to keep the most excessive ornamental space for private use. In order to realize the time-consuming mosaic, which, in our particular case, could be viewed as a weakness of the use of tiles for this project, we flew in an expert and artisan from Spain.

A: 단순하고 견고하며 저렴한 방법으로 표면을 덮을 수 있는 단단한 소재인 것 외에도 복잡한 모자이크를 만드는 능력과 내러티브 및 다각적인 의미를 타일의 가장 큰 장점 중 하나로 간주한다. 우리가 '사무실과 호텔 방' 프로젝트의 화장실 중 하나를 설계할 때 사용하고 싶었던 것이 이 특성이다. 프로젝트의 주요 공간은 흰색에 개방형이고, 중립적이고 유연한 반면, 이 화장실에는 야자수를 묘사한 인상적인 수제 예술 작품이 포함되어 있다. 이는 STAR Strategies + Architecture가 원래 스페인 엘체(Elche)에 있는 공공 광장을 위해 만든 디자인으로, 1 대 8의 축척으로 재현했다. 따라서 아돌프 루스(Adolf Loos)의 찬양자인 의뢰인은 이 가장 과도한 관상용 공간을 사적인 용도에 쓰는 것을 선호했다. 시간이 오래 걸리는 모자이크는 타일의 단점으로 간주 될 수 있지만 우리 프로젝트의 경우 스페인에서 전문가이자 장인을 데려왔다.

GLASS-Q1: Tell us about your favourite project that you used glass in or another architect's work - interior, facade, etc.

A: We recently reached the final phase in an architecture competition for the extension of the District Office in Görlitz, Germany with a project that we called "High and Dry" and in which the use of glass was crucial for its success. Because in order to ensure easy orientation for guests and employees within the entire building complex we created a glazed connecting first floor, providing access to all other buildings of the administrative campus barrier-free, using one of the greatest qualities and strengths of glass: being optically transparent, yet supplying a sufficient division of the inside from the outside of a building. Through this transparent connecting first floor that we created in an elongated administration building inside the block with three wings and three bridges, all the future departments of the enlarged district office will be accessible from the entrance "high and dry".

A: 우리는 최근 독일 괴를리츠(Görlitz)에 있는 지구 사무소를 확장하기 위한 건축 공모전에서 마지막 단계까지 남았었다. 우리는 이 프로젝트를 "높음과 건조함(High and Dry)"이라 불렀고 유리의 사용이 프로젝트의 성공에 결정적인 역할을 했다. 건물 전체 내에서 손님과 직원이 방향을 쉽게 찾을 수 있게 유리로 연결된 1층을 만들었다. 유리의 가장 위대한 성질과 장점을 여기서 이용해 행정 캠퍼스의 다른 모든 건물에 배리어 프리 액세스를 제공할 수 있었다. 시각적으로 투명하되 건물의 외부와 내부 사이에 충분한 분할을 제공하는 점이다. 이 투명하고 연결하는 1층을 통해 우리는 3개의 윙과 3개의 다리가 있는 블록 안의 길게 된 행정 건물을 만들 수 있었고 확장된 지구 사무소의 모든 미래 부서는 입구부터 "높고 건조하게" 접근할 수 있을 것이다.

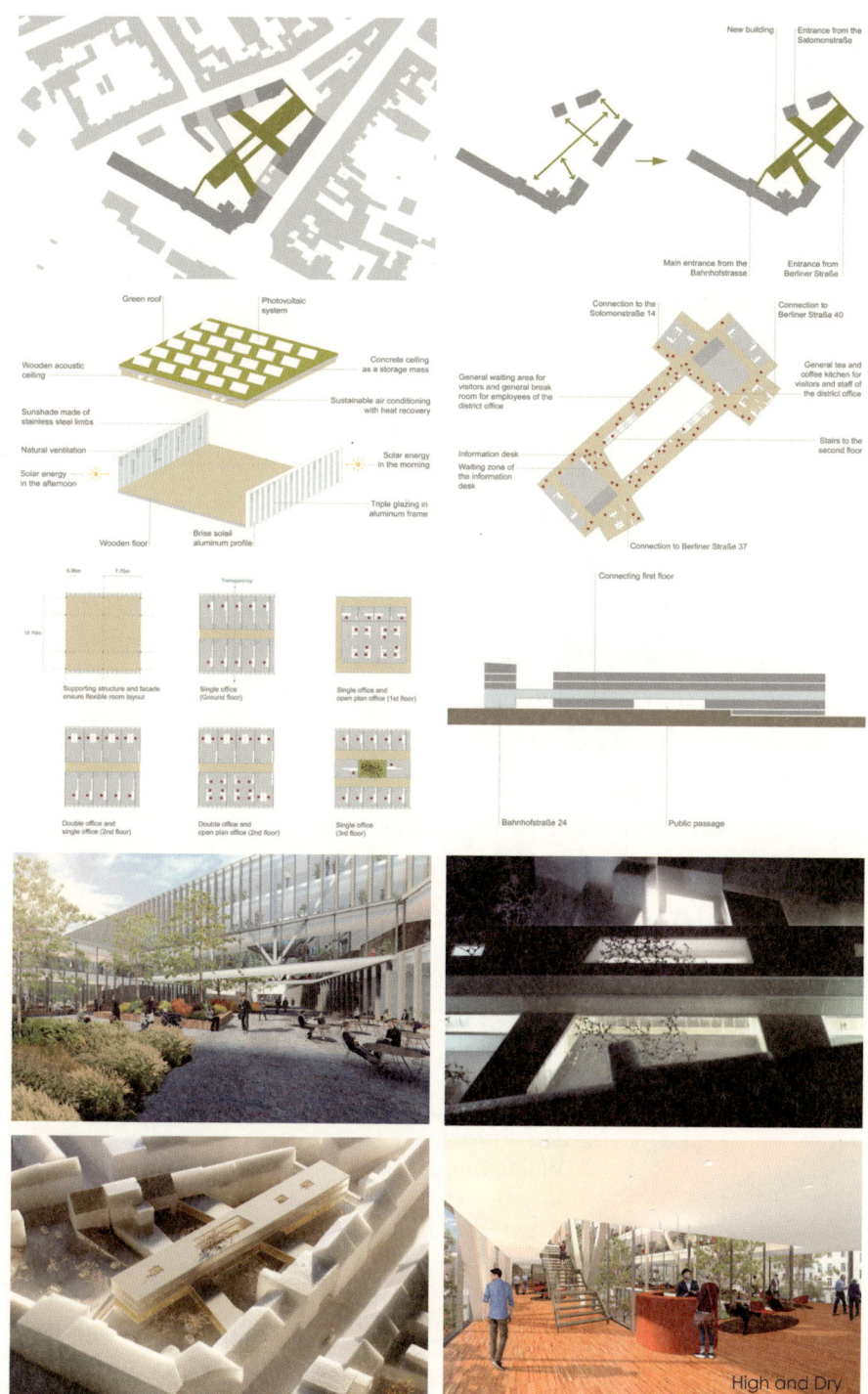

High and Dry

BOARD 103

GLASS-Q2: What are the strengths and weaknesses of glass?

A: When it comes to architecture, one of the considerable weaknesses of glass - next to its porousness that exposes it to the constant risk of cracking – is its ability to transmit light and heat, if too thin and unprotected, that can easily overheat interior spaces during sunny days. To avoid that from happening we proposed to install aluminium profiles as Brise Soleil with a shading effect all over the glass facade of the "High and Dry" project. The use of triple glazing helped keeping the interior spaces in moderate temperatures during the summer months. The proposed highly effective sunscreen made of stainless steel limb-weave also ensured agreeable interior temperatures during hot days. All these measures led to a highly energy efficient administrative building.

A: 건축적인 면에서 끊임없는 균열의 위험에 노출돼있는 다공성 외에, 유리의 상당한 단점 중 하나는 빛과 열을 잘 전달하는 능력이다. 너무 얇거나 보호되지 않으면 맑은 날에 내부 공간을 쉽게 과열할 수 있다. 그런 일이 일어나지 않도록 우리는 "높음과 건조함" 프로젝트의 유리 파사드 전체에 그늘이지는 차양으로서 알루미늄 틀을 설치할 것을 제안했다. 삼중 유리 역시 여름철 내부 공간에 적당한 온도를 유지하는 데 도움이 되었다. 또한 스테인리스 사지 직조로 만든 매우 효과적인 가림막은 더운 날에도 쾌적한 실내 온도를 보장한다. 이러한 모든 방안으로 에너지 효율이 높은 행정 건물을 만들었다.

1. Photovoltaic system
2. Green roof
3. Aluminum sheet
4. Wooden acoustic ceiling
5. Brise soleil aluminum profile
6. Reinforced concrete floor
7. Triple glazing in aluminum frame
8. Sunshade made of stainless steel limbs
9. Wooden floor
10. Sustainable air conditioning with heat recovery

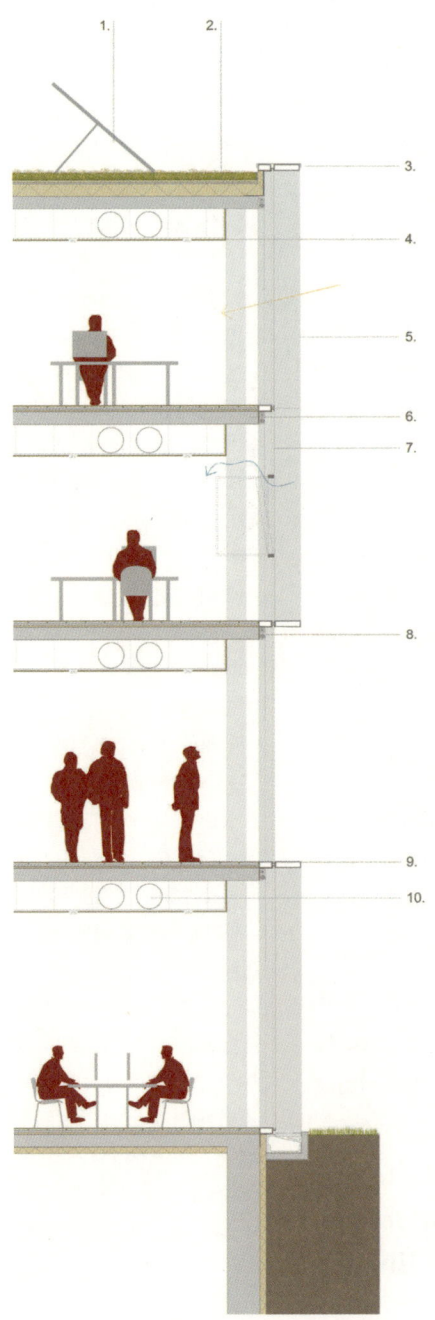

High and Dry

BOARD 105

WOOD-Q1: Tell us about your favourite project that you used wood in or another architect's work - interior, facade, etc.

A: Not so long ago we worked on a design for a tourist centre - a recreational ski complex with 27 ski slopes in the area of Klekovaca Mountain in Bosnia and Herzegovina. When we realized during the design process that for the creation of the ski slopes a large number of trees had to be removed, we decided to re-use them. Thus, we proposed to use the wood from the trees to build the new tourist centre. In that way the wood could be used not only for the structures, but also for the facades of most of the buildings. Accordingly, we called the project "Out of the Woods". With the help of the trees and by using the different geographical features of the site, we suggested to organize the tourist centre in 3 different areas with 3 different layouts and atmospheres embracing the natural beauty of each location: a compact Main Resort Centre next to the skiing plateau and the ski slopes; a Hotel and Sports Complex on the nearly flat area of the site with individual buildings standing free as the trees in the context; and a Climatic

A: 얼마 전 우리는 보스니아 헤르체고비나의 클레코바카산(Klekovaca) 지역에 27개의 스키 슬로프가 있는 레크리에이션 스키 단지인 관광 센터를 디자인했다. 설계 과정에서 스키 슬로프를 만들려면 많은 수의 나무를 제거해야 한다는 점을 깨달았을 때, 자른 나무를 다시 사용하기로 했다. 하여 우리는 사이트의 나무에서 나온 목재를 사용하여 새로운 관광 센터를 건설할 것을 제안했다. 그렇게해서 나무를 구조뿐만 아니라 건물 파사드 대부분에도 사용할 수 있었다. 따라서 우리는 이 프로젝트를 "숲으로(Out of the Woods)"라고 불렀다. 나무의 도움과 사이트의 다양한 지리적 특징을 사용하여 3개의 다른 구역에 각 위치의 자연의 아름다움을 포용하는 3개의 다른 레이아웃과 분위기가 있는 관광 센터를 제안했다. 하나는 스키 고원과 스키 슬로프 옆에 있는 소형 메인 리조트이고, 다른 하나는 사이트 중 거의 평평한 구역에 주변의 나무처럼 자유롭게 서 있는 개별 건물로 구성된 호텔과 스포츠 단지이며,

Health Resort Zone in the pine woods.

세 번째는 소나무 숲에 있는 기후 건강 리조트 구역이다.

WOOD-Q2: What are the strengths and weaknesses of wood?

A: **There are a large number of advantages that made wood one of the most favourite construction materials of humans for thousands of years.** Some of its strengths are without doubt related to its abundant availability, the ability to renew the supply, but also - and in particular - to its longitudinal shape, its toughness but at the same time lightness, and its easy formability and the ease of making joints due to the softness of most of the trees, which makes it a great material to build all kinds of things, such as houses, boats, furniture, to name just a few. However, one of the weaknesses of wood that I would like to point out is that it can not be produced at a particular location in the way bricks, tiles, or glass can be fabricated - trees need to grow for a long time before the wood can be used – and harvest and transport come with a lot of challenges. In the "Out of the Woods" project we wanted to avoid the difficulties related to the transport of wood by using it right at the spot for the construction of the new tourist centre.

A: 수천 년 동안 인간이 가장 좋아하는 건축 자재 중 하나로 꼽힐 만큼 나무에는 많은 장점이 있다. 그 장점 중 하나는 의심할 여지없이 풍부한 가용성, 즉 공급량을 갱신하는 능력이다. 또한 특히 세로 방향의 모양, 강인한 동시에 가벼운 점, 모양을 바꾸기 쉬운 점, 그리고 대부분 나무가 부드러워 이음매를 쉽게 만들 수 있는 점 덕분에 나무는 집, 보트, 가구 등 모든 것을 만드는 데 좋은 재료가 된다. 하지만 지적하고 싶은 나무의 단점 중 하나는 벽돌, 타일 혹은 유리처럼 특정한 장소에서 생산할 수 없다는 것이다. 목재를 사용하기 전에 나무가 오랫동안 자랄 필요가 있고 수확과 운송에는 많은 어려움이 있다. '숲으로' 프로젝트에서 우리는 새로운 관광센터 건설에 바로 그 자리에서 나는 목재를 사용함으로써 목재 운송과 관련된 어려움을 피하고 싶었다.

Carlos Lampreia

Who is ...?

Carlos Lampreia, architect (1990), is architecture design teacher at FAA-Universidade Lusíada de Lisboa since 1994, studied at OPorto Architecture School and at Lisbon Technical University FA-UTL.

Master in architecture theory, 'towards an objective architecture', 2002. Phd about, strategy, site and material, concerning architecture and arts, 'concept site and material, a strategy in architecture and arts, 1960-2000', 2017. His Lisbon based office, carloslampreia[x]arquitectos, works on an experimental way with young architects and students towards architectural materialisation, participating both in international competitions and individual private requests.

Q1: What is material to an architect
    (or to you)?

A: **Material is the result of any transformation we apply into the original matter of our planet.** If in one hand and in this abstract sense, all materials become one, earth matter, on another, there are a lot of different elements on our planet, which we decompose in order to reshape and adapt them to our own habitat.

A: 재료는 우리가 지구상의 물질에 적용하는 모든 변형의 결과이다. 한편으로, 그리고 추상적인 의미에서 모든 재료가 단 하나의 재료이며, 지구의 물질이다. 다른 편으로, 지구에는 다른 요소가 많이 존재하고, 우리가 인간의 서식지에 맞춰 고치고 바꾸기 위해 분해한다.

Q2: Tell us about your favourite
    (or most often used) material and why.

A: In my southern european culture, we are used to solid materials, and we were used to build with earth. so here its very easy for a house to assume some similarities to natural caves. In this way all the materials we generally use to produce walls are stones and bricks that we use to cover with plaster or cement mortar. In this way and in an abstract sense, my favourite material is concrete, because a building made in concrete could easily looks like a direct extrusion of earth.

A: 남부 유럽 문화에서는 순수한 재료가 익숙하고, 흙을 가지고 짓는 데 익숙하다. 그래서 여기는 집이 자연적인 동굴과 비슷하기가 매우 쉽다. 이런 식으로 우리가 일반적으로 벽을 만드는 데 사용하는 모든 재료는 돌과 벽돌이며 석고 또는 시멘트 모르타르로 덮는다. 이런 추상적인 의미에서 내가 가장 좋아하는 재료는 콘크리트이다. 콘크리트로 만든 건물은 쉽게 지구의 직접적인 압출처럼 보일 수 있기 때문이다.

Q3: When do you decide the material during the design process and what is your criteria? (e.g. budget, client's preference, design concept, climate, etc.)

A: All decisions in terms of materials should be related to a general design concept, a strategy where all the others aspects, like site particularities, budget, climate or preferences should be integrated.

A: 재료 측면에서 모든 결정은 사이트의 특수성, 예산, 기후 또는 환경 설정과 같은 다른 모든 측면을 통합해야 하는 전략인 전체적인 디자인 콘셉트와 관련되어야 한다.

Q4: Tell us about the materials you are interested in or want to use in your projects right now.

A: Right now, because of the global production crisis, i think there is a tendency to decrease industrial materials and increasing the crafts and low tech materials. In this sense **I am now interest in materials concerning site and memory, that might be an answer to every local issues around the world.**

A: 현재 세계적인 생산 위기 때문에 산업 재료를 줄이고 공예품과 수준이 낮은 기술의 재료를 늘리는 경향이 있다고 생각한다. 이런 의미에서 **나는 현재 사이트와 기억에 관한 재료에 관심이 있으며, 이는 전 세계의 모든 지역 문제에 대한 해답일 수 있다.**

Carlos Lampreia

BRICK-Q1: Tell us about your favorite project that you used brick in or another architect's work - interior, facade, etc.

A: One of my favourites houses made in brick, is the experimental Muuratsalo House by Aalto in Finland, where in the patio we can watch a range of different textures made by bricks working together against the external white painted bricks that helps defining the edge of the building.

A: 내가 가장 좋아하는 벽돌로 된 집 중 하나는 핀란드에 있는 알토(Aalto)의 무라찰로 실험집(Muuratsalo Experimental House)이다. 테라스에서 벽돌이 만드는 다양한 질감을 볼 수 있고 이는 외부의 흰색 벽돌에 맞서 건물의 윤곽을 더 뚜렷이 보이게 한다.

Muuratsalo Experimental House
©Timothy Brown

BRICK-Q2: What are the strengths and weaknesses of brick?

A: The strength of brick to me is about surface and order in construction, and also because when someone put them

A: 벽돌의 장점은 건설할 때의 표면과 질서이고 누군가 다 쌓으면 얻을 수 있는 마감된 표면과 질감이다.

112 Brick, Brick! What do you want to be?

together gets a finished surface and a texture.
Has Louis Kahn once said a brick 'likes an arch', and in that sense it's a material that works well producing massive walls, but needing arches or other types of substructures in order to achieve voids between those walls.

루이스 칸(Louis Kahn)은 벽돌이 '아치를 좋아한다'고 말했다. 그런 의미에서 거대한 벽을 잘 만드는 재료이지만 벽 사이의 공간을 달성하기 위해 아치 또는 다른 유형의 하부구조를 필요로 한다.

TILE-Q1: Tell us about your favourite project that you used tile in or another architect's work - interior, facade, etc.

A: Tiles are a very popular material in Portugal, and i like to use them as a panel to reflect light and protect the walls from water like in the Chiado Building in Lisbon by Alvaro Siza, where soft coloured tiles in the facades bring light and sun reflections from the river to the streets around.

A: 타일은 포르투갈에서 매우 인기 있는 재료이다. 알바로 시자(Alvaro Siza)가 지은 리스본의 치아도(Chiado) 건물과 같이 빛을 반사하고 물에서 벽을 보호하기 위해 타일을 패널로 사용하는 것을 좋아한다. 파사드에 있는 부드러운 색의 타일은 빛과 강에서 반사되는 햇빛을 건물 주변의 길로 불러온다.

TILE-Q2: What are the strengths and weaknesses of tile?

A: Tiles are a coating material, usually used to cover walls or floors inside and outside while defending them from water.

A: 타일은 코팅된 재료로, 일반적으로 벽이나 바닥을 안팎으로 덮고 물에서 보호하는 데

Since its a ceramic product it can improve the light and receive a full range of colour and textures. The weakness about them is firstly their thickness, and secondly that they can became fragile with the use and constructive tensions.

사용된다. 세라믹 제품이기 때문에 빛을 개선할 수 있고 모든 색상과 질감이 가능하다. 단점은 첫째로 그 두께이며, 둘째로 사용과 건설적인 장력으로 깨지기 쉽다는 점이다.

Chiado Building in Lisbon ©Carlos Lampreia

GLASS-Q1: Tell us about your favourite project that you used glass in or another architect's work - interior, facade, etc.

A: A good example of using glass as a transparent wall is the deutsch pavillion in barcelona by Mies van der Rohe, where we can feel the abilities of the material as a filter towards the exterior, wherever is a square or nature like in the patio. In both situation glass has the ability to provide a very strong feeling of continuity but also a very cosy internal space.

A: 유리를 투명한 벽으로 사용하는 좋은 예는 미즈 반 데어 로에(Mies van der Rohe)가 바르셀로나에 지은 독일 파빌리온(Deutsch Pavilion)이다. 이곳에서 안뜰과 같은 정사각형이나 자연이 있는 곳이면 어디에서나 외부를 향한 필터로서 유리의 능력을 느낄 수 있다. 두 경우 모두 유리는 매우 강한 연속적인 느낌을 줄 뿐만 아니라 매우 아늑한 내부 공간을 만든다.

Deutsch Pavilion ©Alice Wiegand

GLASS-Q2: What are the strengths and weaknesses of glass?

A: Glass is a marvelous material, because in strict sense we can get a transparent wall using it. But sometimes we commit the error of looking at glass as an invisible material, and that's a mistake. Glass its the most visible of all materials, because it is a perfect and shining highly proccessed material. The bigger weakness of glass is its fragility and dependence on other substructures that are required while using it.

A: 유리는 놀라운 물질이다. 왜냐하면 유리를 사용하여 완전하게 투명한 벽을 지을 수 있기 때문이다. 하지만 가끔 유리를 보이지 않는 물질로 생각하는 오류를 저지르는 데 그것은 실수다. 유리는 완벽하고 빛나는 매우 가공된 물질이기 때문에 모든 재료 중에서 가장 눈에 띈다. 더 큰 단점은 그 취약성과 유리를 사용하는 데 필요한 하부 구조에 대한 의존성이다.

WOOD-Q1: Tell us about your favourite project that you used wood in or another architect's work - interior, facade, etc.

A: In Fisher house, a wood house made by Louis Khan, i found amazing the way wood creates a corner space at the living room, joining the external cover with a window, a bench, and some furniture pieces, all made by the same type of wood in the same piece of design.

A: 루이스 칸이 지은 목재집 피셔 하우스(Fisher House)에서 거실에 나무로 모서리 공간을 만들어 내는 방식이 놀라웠다. 외부 커버를 창문, 벤치와 다른 가구가 연결되어있는데 모두 같은 종류의 나무로 만들어진 하나의 디자인이다.

WOOD-Q2: What are the strengths and weaknesses of wood?

A: **Wood and earth were the first materials mankind used to provide shelter.** With wood we can build almost everything from an house to furniture. In this sense having wood is like have a friend at home, since its a very organic and lived material that moves, sounds and smell, we may even look at it as an artificial tree. The only problem relies in the fact that it is impossible to have an entire city build on wood because we also need forests. This undeniable reality makes wood an expensive and rare material that we should use with care.

A: 나무와 흙은 인류가 피난처를 짓기 위해 사용한 최초의 재료이다. 나무로 우리는 집에서 가구까지 거의 모든 것을 만들 수 있다. 이런 의미에서 집에 나무가 있는 것은 마치 친구를 집에 둔 것 같다. 목재는 매우 유기적이고 살아있는 재료이며, 움직이고, 소리와 향기가 있기 때문에 인공 나무로 생각할 수도 있다. 유일한 문제는 숲이 필요하기 때문에 도시 전체를 나무로 만드는 것이 불가능하다는 점이다. 이 부정할 수 없는 현실은 나무를 조심스럽게 사용해야 하는 비싸고 희귀한 재료로 만든다.

"MATERIAL IS THE RESULT OF ANY TRANSFORMATION WE APPLY INTO THE ORIGINAL MATTER OF OUR PLANET."

Carlos Lampreia

Casanova+Hernandez
Architects

Who is ...?

Casanova+Hernandez, founded in 2001 by Helena Casanova and Jesus Hernandez, is a design and research studio based in Rotterdam. It focuses on rethinking and designing our urban habitat in order to create vibrant cities while promoting environmental and social sustainability.

Working with an interdisciplinary team and with experience developing projects in very different cultural contexts in Europe, South America and Asia, the office has expanded its capabilities and its international network through close and fruitful collaboration with experts in different continents.

Q1: What is material to an architect (or to you)?

A: **Material is an important part of the architectural concept.**
The material is directly responsible for the expression of the building. By choosing one material or other, the architect decides if the building is fully integrated into its surroundings, if it stands isolated from it, if it creates a dialogue with certain areas of it. By taking a decision over materials the architect decides if the building will have a special story to tell or will stay anonymous and neutral, if it will become an icon or if it will be mimicked with what is around it.

On the urban scale, both, volume and material, communicate a message to the passers-by. A transparent building with a big opening on the ground floor can become an inviting piece of architecture, a concrete box with little or no openings creates distance with its surroundings.

On the smaller scale of the building itself, materials also have a big impact in the way users experience the building as a warm, cold, inviting or unwelcoming space.

When choosing a material, the architect also enters a wide world of possibilities when considering the different properties

A: 재료는 건축 디자인 컨셉에서 중요한 부분이다.
재료는 건물의 표현을 직접 책임진다. 하나의 재료를 선택함으로써 건축가는 건물이 주변 환경에 완전히 통합될지, 격리될지, 특정 구역과 조화를 이룰지 여부를 결정한다. 재료에 대한 결정을 내린 후 건축가는 건물이 익명의 중립적인 이야기를 할 것인지, 상징이 될 것인지, 아니면 주변 환경을 모방할 것인지를 결정한다. 도시의 축척으로 볼 때 용적과 재료는 모두 통행인에게 메시지를 전달한다. 1층에 큰 개구부가 있는 투명한 건물은 우호적인 건축물이 될 수 있으며, 개구부가 거의 없거나 전혀 없는 콘크리트 상자는 주변 환경과 거리를 만든다. 좀 더 작은 건물의 축척으로 볼 때 재료는 사용자가 따뜻하든, 차갑든, 매력적이든, 안락해 보이지 못하든, 건물을 경험하는 방식에 큰 영향을 미친다. 재료를 선택할 때 건축가는 각 재료의 다른 특성을 고려하며 다양한 가능성의 세계에 빠진다. 색상, 질감 혹은 냄새까지 나중에 감각이 건물을 인식하는 방식을 정한다.

of each material. Color, texture and even smell will define later the way the building is perceived by human senses.

At our office, the selection of material plays an essential role since the very first step of the design process. After analyzing the location of the project, we immediately start thinking about which role our building should play in its surroundings. Which message should our building transmit? Should it be more open or guarantee more privacy, should it be more massive and silent or should it become an icon in the area?

Considering the perception of the material under the phenomenological dimension of architecture, we decide both, volume and material simultaneously at the beginning of the design process. We also give special attention to this perception of our buildings following three different scales, the urban scale, the scale of the block and the human scale of the interiors.

We normally design considering materials that allow us to play and manipulate the perception of the building and its direct surroundings.

우리 사무실에서 재료의 선택은 디자인 과정의 첫 단계부터 필수적인 역할을 한다. 프로젝트의 위치를 분석한 후, 우리 건물이 주변에서 어떤 역할을 해야 하는지 곧바로 생각한다. 우리 건물은 어떤 메시지를 전해야 하는가? 더 개방되어야 하는가, 사생활을 더 보장해야 하는가? 더 거대하고 고요해야 하는가, 아니면 이 지역의 상징이 되어야 하는가? 우리는 건축의 현상학적 차원에서 재료의 인식을 고려하여, 디자인 과정의 시작 부분에서 용적과 재료를 동시에 결정한다. 우리는 또한 세 가지 다른 축척, 즉 도시, 블록 그리고 실내의 인간 축척에 따른 우리 건물의 이러한 인식에 특별한 주의를 기울인다.
우리는 일반적으로 건물과 그 직접적인 환경에 대한 인식을 실험해보고 다룰 수 있는 재료를 고려하여 설계한다.

Q2: Tell us about your favourite (or most often used) material and why.

A: **Without any doubt, our favorite material is glass.**
Glass is a very versatile material. Glass allows us to design thinking of a wide range of possibilities to create different mechanisms to control the way our buildings are perceived.
Glass can be transparent, but it can also be translucent or even opaque.
In 1980 Paul Virilio stated in his book The Aesthetics of Disappearance that 'after the age of architecture-sculpture we are now in the time of cinematographic fictitiousness; literally as well as figuratively, from now on architecture is only a movie'. The architecture of disappearance makes sense if we interpret the term disappearance not as the absence of materiality, but as a lack of visual appearance or at least as an ambiguous appearance.
The particular properties of glass such as transparency, translucency, and reflection allow glazed facades in combination with natural and artificial light to play an effective role in creating invisible skins capable of provoking an ambiguous perception of the building. Sometimes

A: 우리가 가장 좋아하는 재료는 두말할 것도 없이 유리이다.
유리는 매우 다재다능한 물질이다. 우리가 건물을 인식함에 있어 다양한 방법으로 접근할 수 있게 하는 넓은 범위의 디자인 사고(design thinking)를 가능하게 한다.
유리는 투명 할 수 있지만 반투명하거나 불투명할 수도 있다.
1980년 폴 비릴리오(Paul Virilio)는 자신의 저서 『소실의 미학』에서 "조각-건축 시대 이후 우리는 이제 영화적 허구의 시대에 접어들었다. 문자 그대로뿐만 아니라 비유적으로도, 건축은 이제부터 단지 영화일 뿐이다"라고 말했다. 소실이라는 용어를 물질성이 없는 것이 아니라 시각적 외관이 부족하거나 적어도 모호한 모습이라고 해석하면 소실의 건축이란 말이 이해가 된다.
투명성, 반투명성, 반사와 같은 유리의 특성은 자연광 및 조명과 함께 유리 파사드가 건물에 대한 모호한 인식을 유발하는 보이지 않는 표면을 만드는 데 효과적인 역할을 한다. 때로 이 모호한 인식은 건물과 유리 외관에 반사된 주변 환경에 대한 지각이 합쳐질 때 발생한다. 다른 때에는 반투명한 외관에 자연광이 비추어 건물

The reflection on the glass of the abundant nature around the building merges with the printed vegetation on the glass.

유리에 반사된 건물 주변의 풍부한 자연과 유리에 인쇄된 식물이 병합된다.

©Christian Richters

this ambiguous perception results when the perception of the building melts with the perception of the reflection of the surroundings on its glazed facade; at other times it results when the incidence of natural light on the translucent facade produces the optical dissolution of the contour of the building.

Playing with glasses with different levels of transparency in some of our projects, we have been able to capture the deepness of light in all its richness, varying over time according to different hours of the day and different seasons of the year. Translucent glass does not only reflect natural light differently depending on the passing of time, it also projects interior lights towards the exterior provoking the creation of silhouettes with different degrees of blackness, which exposes the building as a sort of theatrical stage towards the outside world, establishing a poetic dialogue with the surroundings. Light and shadows, stillness and movement communicate a story about the interior life of the building.

Glass can also present colors and printed patterns, which adds interesting layers to its value as an interface between the interior life of a building and its public image.

윤곽의 시각적 용해가 생긴다. 우리는 여러 프로젝트 중 일부에서 투명성이 다른 유리를 가지고 실험해보면서 시간과 계절이 변함에 따라 변화하는 빛의 풍부함과 깊이를 포착할 수 있었다. 반투명 유리는 시간의 경과에 따라 자연 채광을 다르게 반사할 뿐만 아니라 외부를 향하는 내부 조명을 투사한다. 이는 다양한 정도를 지닌 흑색의 실루엣을 유발하고 건물은 일종의 연극 무대가 되어, 주변 환경과 시적인 대화가 이루어진다.

빛과 그림자, 고요함과 움직임은 건물의 내부에 관한 이야기를 전한다.

또한 유리에는 색상과 인쇄된 패턴을 쓸 수 있으며, 이는 건물의 내부 생활과 공공 이미지 사이의 접점으로서 유리의 가치에 흥미로운 의미를 추가한다. 유리의 색, 인쇄된 모티브의 주제 및 규모 같은 요소를 조절함으로써 우리는 다른 축척에서 다른 의미를 가진 '말하는' 파사드를 만들었다. 네덜란드 아펠도른(Apeldoorn)에서 진행 중인 공동 주거 프로젝트는 공원 앞에 위치하기 때문에 주변의 녹색 환경에 통합되어야 했다. 물리적으로나 지각적으로나 건물과 주변 환경을 통합하기 위해 은행나무 모티브가 인쇄된 파사드를 특별히 디자인했다. 이 파사드는 유리 파사드와 그 위에 인쇄된 식물 모티브에 여

By controlling aspects such as the color of the glass, the subject and the scale of the printed motives we have created 'talking' facades with different readings on different scales. In our project in Apeldoorn, the Netherlands, a collective housing project had to be integrated into a green environment, in front of a park.

In order to integrate it, both, physically and in terms of perception, we designed specifically a facade with a printed motif of Ginkgo Biloba tree.

This facade is an experiment with the perception of disappearance through the use of several levels of transparency in the glazed facade and the vegetal motives printed on it. The reflection on the glass of the abundant nature around the building merges with the printed vegetation on the glass. The talking skin works as a device that confuses our senses, minimizes the physical presence of the building and renders the comprehension of its ambiguous image more difficult.

The printed glazed skin of the Ginkgo project observed from a far distance visually integrates the building with the greenery of the park. In a closer view of its facade, the observer clearly perceives the realistic image of branches and leaves printed on the glass, which plays with

러 수준의 투명성을 사용하여 소실에 대한 인식을 실험한 것이다. 유리에 반사된 건물 주변의 풍부한 자연과 유리에 인쇄된 식물이 병합된다. '말하는' 파사드는 감각을 혼란스럽게 하고, 건물의 물리적 존재감을 최소화하며, 모호한 이미지의 이해를 더욱 어렵게 만든다. 멀리서 보면 은행나무 프로젝트의 인쇄된 유리 파사드는 건물을 공원의 녹지와 시각적으로 통합한다. 파사드를 더 자세히 보면, 관찰자는 유리에 인쇄된 가지와 잎의 현실적인 이미지를 분명히 인식한다. 이 이미지는 현실과 환상, 자연과 인공물, 예술과 건축을 나누는 의식과 무의식 사이를 활용한다. 파사드를 자세히 살펴보면, 관찰자는 나뭇가지를 따라 각기 다른 곳에 위장한 작은 곤충과 같은 작은 디테일의 우주를 발견한다. 이는 건물을 놀랍고, 생생하고, 예측할 수 없게 만든다.

the conscious and unconscious line that divides reality and illusion, nature and artifice, and art and architecture. In a close look at the facade, the observer discovers a universe of small details such as small insects camouflaged at different points along the branches, which makes the building surprising, alive and unpredictable.

Q3: When do you decide the material during the design process and what is your criteria? (e.g. budget, client's preference, design concept, climate, etc.)

A: **We decide the material at the very beginning of the design process**, after deciding which kind of role our building should play in its surroundings, considering not only its physical location but also the cultural meaning of the place, its history and the non-verbal layers of meaning associated to the commission. We start a design process analyzing all the psychological and sociological aspects relevant for the commission within its specific cultural context, together with the physical relevant aspects of the given location, such as the climate conditions, or the given budget.

A: 우리는 디자인 과정의 시작 부분에서 재료를 결정한다. 물리적 위치뿐만 아니라 우리 프로젝트와 연관된 그 장소의 문화적 의미, 역사 및 비언어적 의미 등을 고려하여 우리 건물이 주변에서 어떤 역할을 해야 하는지 결정한 후에 재료를 결정한다. 우리는 기후 조건이나 예산 같은 주어진 위치의 물리적인 면과 특정 문화적 맥락에서 프로젝트에 관련된 모든 심리적 및 사회학적 측면을 함께 분석하며 디자인을 시작한다.

Q4: What are some architectural projects that inspired you regarding brick, tile, wood and/or glass? And why?

A: Brick-On the one hand, the refined works in terracotta that are characteristics of the mosques in Bogra or the temples in Puthia, Bangladesh, have shown us the enormous possibilities of this material. Though such textures and refinements are not directly readable in our architecture, those works are for us referential when we talk about the beauty of brickwork. Some Alvar Aalto's works such as the experimental house in Muuratsalo belong to this category of architectural works in brick that we admire.

On the other hand, the traditional gray bricks used in the hutongs in Beijing, with their proportions and different gray tones and textures, have become as well referential in our work, due to their elegant, sober beauty. During our visit to Caochangdi district in Beijing, we were quite impressed by how Ai Wei Wei used this kind of brick in a contemporary way.

Tiles-Our work with tiles to create colorful landscapes and architectural objects is inspired by the work of the modernist Catalan architects from the

A: 벽돌 한편으로, 방글라데시 보그라(Bogura)의 이슬람 사원이나 푸티아(Pythia)에 있는 사원의 특징인 테라코타로 된 세련된 작품은 우리에게 이 재료의 엄청난 가능성을 보여준다. 우리 건축에서는 이러한 질감과 세련미를 직접 읽을 수는 없지만, 벽돌 작업의 아름다움에 관해 이야기할 때 우리에게 참고가 된다. 무라차로(Muuratsalo)의 실험용 집과 같은 알바르 알토(Alvar Aalto)의 작품 중 일부는 우리가 감탄하는 벽돌로 된 건축 작품의 범주에 속한다. 다른 편으로는 베이징의 후통에서 사용된 비율과 다양한 색조 및 질감의 전통적인 회색 벽돌은 우아하고 냉담한 아름다움이 우리에게 참고가 되었다. 베이징의 차오창띠 지역을 방문했을 때, 우리는 아이 웨이 웨이가 이런 종류의 벽돌을 현대적으로 어떻게 사용했는지에 깊은 인상을 받았다.

타일 다채로운 풍경과 건축 물체를 만들기 위해 타일을 쓴 우리의 작품은 19세기와 20세

Brick example

Casanova+Hernandez Architects 127

19th century and the beginning of the 20th century. By using the trencadís technique, characteristic of Gaudi's work, that is based on placing broken ceramic pieces on a bed of cement, we have been able to cover complex surfaces, both, horizontal and vertical, which has allowed us to build interventions characterized by the continuity of surfaces and material. Thanks to the possibility of combining different colors and patterns of broken tiles, designing with tiles opens up a wide range of possibilities regarding the expression of buildings and landscapes, which gives us a lot of freedom as designers.

Glass-When in 2003 we were invited to make a proposal for the new Tittot Art Glass Museum in Taipei, Taiwan, to exhibit its splendid collection of contemporary art in glass, we had the opportunity of getting in touch with some of the artists of this collection. Thanks to this direct contact with artists who worked with glass and to our own research about glass properties and methods and techniques used to create different degrees of transparency, prints on glass, etc. we developed wide knowledge on this material. Dan Graham's work, which

기 초의 모더니스트 카탈루냐 건축가에서 영감을 얻었다. 가우디 작품의 특징이며 깨진 세라믹 조각을 시멘트에 놓는 것이 기반인 트렌카디스(trencadís) 기법을 이용해서 우리는 수평과 수직의 복잡한 표면을 덮을 수 있었고, 이로 표면과 물질의 연속성을 특징으로 하는 인터벤션을 지을 수 있었다. 깨진 타일의 다양한 색상과 패턴을 결합할 수 있는 점 덕분에 타일로 디자인하는 것은 건물과 풍경의 표현에 관한 다양한 가능성을 열어 디자이너에게 많은 자유가 생긴다.

유리
2003년 대만 타이베이에 있는 새로운 티토 아트 글래스 박물관(Tittot Art Glass Museum)에 현대 유리 미술의 훌륭한 컬렉션을 전시하기 위한 디자인을 내도록 초청받았을 때, 우리는 이 컬렉션의 일부 예술가와 연락할 기회가 있었다. 유리로 작업한 예술가와의 직접적인 대화와 다양한 투명성, 패턴 인쇄 등을 만드는 데 사용되는 유리의 특성, 방법 및 기술에 대한 연구 덕분에 우리는 이 재료에 대한 폭넓은 지식을 쌓았다. 건축적 차원을 가진 댄 그레이엄

the new Tittot Art Glass Museum in Taipei, Taiwan

has an architectural dimension, has also inspired our work.
Glass is our favorite material to work with. In very different projects and different scales we have used glass as a

(Dan Graham)의 작품 또한 우리에게 영감을 주었다.

유리는 우리가 사용하기 가장 좋아하는 재료이다. 매우 다른 프로젝트와 다양한 축척에

tool to blend architecture with nature or as a means to create atmospheric spaces in our buildings and to explore the poetics of space through the conscious capture of the reflections of both, natural and artificial light, as well as of silhouettes in the facades of our buildings.

Wood-Our inspiration when working with wood doesn't come from architecture works but from the nests made by different species of birds. In our office, we have studied the shapes of birds' nests and have tried to catch the spirit of such interesting objects, while rationalizing them into architectural ones.

By getting inspired by natural objects instead of by buildings, we are freer as designers to create special interventions where unusual shapes, wooden textures, and colors intermingle and provide a work of architecture with an artistic

서 우리는 건축을 자연과 조화시키는 도구로, 건물에 분위기 있는 공간을 만들 방법으로, 그리고 자연과 인공 빛의 반사뿐만 아니라 건물 파사드의 실루엣을 의식적으로 포착하여 공간의 시학을 탐구하는 수단으로 유리를 사용했다.

나무
나무로 작업할 때의 우리의 영감은 건축 작업에서 나온 것이 아니라 다른 종의 새들이 만든 둥지에서 나온다. 우리 사무실은 새 둥지의 모양을 연구했으며 이 흥미로운 물체의 정신을 포착하고 건축학적으로 합리화하려 노력했다. 건물 대신 자연에서 영감을 얻음으로써 우리는 특이한 모양, 나무 질감 및 색상이 섞이고 예술적 차원의 건축작품의 특별한 인터벤션을 만드는 데 디자이너로서 더 자유롭다. 우리는 모든

Natural Object

the Nest

dimension. We pay special attention to the phenomenological dimension of architecture in all of our works and by working with wood and the wide range of possibilities it allows, we can awake feelings and emotions that are not possible to raise by using other materials.

As Gaston Bachelard in his book The Poetics of Space mentions,
With nests ... we shall find a whole series of images ...that brings out the primitiveness in us.

We play in our works in wood with those resonances of the nest as a primal shelter, where humans feel protected and safe, where life takes place quietly and in harmony with nature.

작품에서 건축의 현상학적 차원에 특별한 주의를 기울인다. 나무와 함께 작업하고 그것이 허용하는 다양한 가능성을 통해 다른 재료로는 만들 수 없는 느낌과 감정을 깨울 수 있다.

가스통 바슐라르(Gaston Bachelard)가 쓴 책 "공간의 시학"에서 언급했듯이 **"둥지에서…. 우리에게 원시성을 불러일으키는…. 일련의 이미지를 발견할 것이다"**.

우리는 나무로 된 작업에서 인간이 보호받고 안전하다고 느끼고 생명이 조용히 자연과 조화를 이루는 곳인 태고의 주거지로서 둥지의 공명을 활용한다.

Q5: Tell us about the materials you are interested in or want to use in your projects right now.

A: The material we are making research on right now at our office is wood. More and more, our office is invited to conceive and develop big plans consisting of networks of interventions in protected nature reserves or in natural unexploited areas where we build interventions exploring the limits between architecture, landscape architecture, and art. For this kind of projects, where protecting the existing nature is essential, we work with wooden objects that are used as information points on the one hand, but that, on the other hand, work as pure sculptural objects inserted in nature. The main aim of this kind of projects is to create eco-friendly networks by building isolated interventions that are at the same time conceptually linked and coordinated with other interventions in the area. In that way, all interventions together create tourist routes where visitors are informed about the history of the place, its ecosystem, the traditions and craftsmanship skills present in the area, etc.

A: 지금 우리 사무실에서 연구하고 있는 재료는 나무이다. 우리 사무실은 점점 더 자연보호 구역이나 개발되지 않은 녹지에 인터벤션 네트워크로 구성된 대규모 계획을 고안하고 개발하도록 초청받고 있으며 건축, 조경 건축 및 예술 사이의 한계를 탐구하는 인터벤션을 짓는다. 기존의 자연을 보호하는 것이 필수인 이런 종류의 프로젝트의 경우, 우리는 한편으로는 정보 포인트로, 다른 한편으로는 자연에 삽입된 순수한 조각물인 나무 물체를 만든다. 이러한 종류의 프로젝트의 주요 목적은 따로 떨어진 인터벤션을 짓되 그 주변의 다른 인터벤션과 개념적으로 연결되고 조화로운 친환경적인 네트워크를 만드는 것이다. 그런 식으로 모든 인터벤션은 방문객이 그 장소의 역사, 생태계, 지역의 전통과 장인 기술 등에 대해 알게 되는 관광 루트를 만든다.

For the Frampton Marsh area, the biggest marsh in UK, we are going to build a wooden intervention called The Tower, where visitors will be able to watch birds and the marsh from an advantaged point of view, but where, at the same time, they will be informed about the outstanding works of art inspired by this marsh in the past made by artists such as Turner.

For this project, we have made research on the existing systems to protect the banks of the rivers in the area with big hardwood pieces, resistant to very harsh climate conditions. The properties of these pieces, such as its hardness, color, smell and texture, will help us to define a sustainable object integrated into the landscape, for visitors to rest, get sheltered and get inspired by nature and by the history of the place.

The Tower

BRICK-Q1: Tell us about your favorite project that you used brick in or another architect's work - interior, facade, etc.

A: In our project for a collective housing block in Blaricum, the Netherlands, we used black brick as a means to create a cubic pure volume used as a canvas on which to perforate a series of varied openings framed with a white aluminum strip in order to give importance to the void instead of to the architectural mass.

The 'extraction of mass' in this project from the ideal black brick prismatic

A: 네덜란드 블라리쿰(Blaricum)에 있는 공동 주거 단지 프로젝트에서 우리는 건축적 매스보다 보이드(void)에 중요성을 부여했다. 가느다랗고 하얀 알루미늄 테가 둘린 다양한 구멍을 뚫을 캔버스로써 검은 벽돌로 순수한 용적을 만들었다.

이 프로젝트의 '매스 추출'은 검은 벽돌로 된 이상적인

Black & White Twins ©Christian Richters

volume, especially when it is extracted from the four corners of the building, reduces its massive appearance. The white color in the voids contrasts with the dark color of the brick facade to create what, according to the reification property of the gestalt theory, is called a subjective or illusory contour. As happens in many carved wooden sculptures created by Barbara Hepworth during the 1940s, color helps to render visible the invisible and the void becomes an important constructive element of the work. The black brick reinforces the appearance of the outer skin of the building and the white frames underline the presence of the voids perforated into it.

BRICK-Q2: What are the strengths and weaknesses of brick?

A: In our opinion, **brick is a good means to create a neutral base on top of which other materials can get the predominant expressive role.**
Brick can be also an elegant, silent material to express serenity and neutrality in architecture if its quality is good and its texture, proportions, and color are carefully chosen.

The weakness of brick is that it can also be understood as a quite vulgar material if it is not chosen and used carefully.

속한 재료로 보일 수 있다는 점이다.

TILE-Q1: Tell us about your favourite project that you used tile in or another architect's work - interior, facade, etc.

A: For our project in Jinzhou, China, we have used broken tiles to create a hybrid landscape consisting of a park and a museum where the limits between landscape and architecture are erased. By using the same broken tiles of 4 different colors to create the pavement and the benches of the park on the one hand, and the facades and the roof of the museum on the other, a three-dimensional landscape is created, understood as a complete work of art where colors and reflections unify the complex.

In this case, the material, its properties and its possibilities of use, have been at the very core of the entire design concept. The commission required from all invited designers to reflect on their own national identity. Instead of entering the narrow field of national identities, we preferred

A: 중국 진저우의 프로젝트를 위해 우리는 깨진 타일을 사용하여 풍경과 건축 사이의 경계가 지워지는 공원과 박물관으로 구성된 하이브리드 풍경을 만들었다. 4가지 색상의 깨진 타일을 사용하여 한편으로는 길과 공원 벤치를, 다른 한편으로는 박물관의 정면과 지붕을 만들어 3차원의 풍경을 만들었다. 타일의 색상과 반사가 이 모두를 통합하여 프로젝트는 완전한 예술 작품이 된다.

이 경우 재료, 그 특성 및 사용 가능성은 전체 디자인 컨셉의 매우 중요한 핵심이었다. 이 프로젝트는 모든 초청 디자이너가 각자의 국가 정체성을 반영해야 했다. 우리는 국가 정체성이라는 좁은 영역에 들어가는 대신, 이미 수 세기 동안 이루어져 온 서구와 동양 간의 모든 상업 및 문화 관계를 대표하는 현상인 문화적 하이브

Ceramic Museum and Mosaic Park ©Mac Millan

Casanova+Hernandez Architects

to make a reflection on the concept of cultural hybridization, a phenomenon representative of all commercial and cultural relations between West and East during centuries already. The development and use of ceramics as mosaics in the West on the one hand and the development of the crackled glaze of the Chinese porcelain on the other offered us an interesting conceptual starting point to make research on and to base our design. The material defined the design since the beginning of the design process.

리드화의 개념을 반영하는 것을 선호했다. 서양의 모자이크로서 도자기의 개발 및 사용과, 중국 도자기에서 금이 간 유약의 개발은 우리의 조사를 시작하고 디자인의 기초를 둘 흥미로운 개념적 출발점을 제공했다. 여기서 재료는 디자인 과정의 시작부터 전체적인 디자인을 정의했다.

TILE-Q2: What are the strengths and weaknesses of tile?

A: By using broken tiles, in the same way that Gaudi and the Catalan modernist architects did, to cover pavements, facades, and urban elements, we have been able to create unique hybrid landscapes with complex geometries in which all horizontal, diagonal and vertical planes work synchronically as a stage where urban life occurs. By breaking the tiles into small pieces, they can be adapted easily to almost any surface, which gives the designer a lot of freedom of

A: 가우디(Gaudi)와 스페인 카탈루냐(Catalonia)의 모더니스트 건축가들이 그랬던 것처럼, 우리는 도로, 파사드, 도시 요소를 덮기 위해 부서진 타일을 사용했다. 덕분에 모든 수평, 대각선 및 수직면이 도시 생활이 일어나는 무대로 동기화하는 복잡한 기하학의 독특한 하이브리드 풍경을 만들 수 있었다. 타일을 작은 조각으로 부수면 거의 모든 표면에 쉽게 적용할 수 있어 디자이너에게 표현의 자유가 많이 부여된다.

Gaudi's work

expression.

The versatility in colors and patterns of the broken tiles provides as well a high degree of freedom of design and of chromatic richness. This has allowed us to play with the perception of the whole by defining the relation between the different parts of a design in terms of colors and patterns, thus providing the opportunity of creating perspective effects and rich views throughout buildings and landscapes.

The weakness of tiles is that they can be broken easily under certain circumstances, but if they are applied already broken on purpose, this point becomes irrelevant.

깨진 타일의 다양한 색상과 패턴은 디자인할 때 많은 자유와 풍부한 색채를 제공한다. 덕분에 우리는 색상과 패턴면에서 각기 다른 디자인 요소 사이의 관계를 정의하여 건물 전체에 대한 인식으로 여러가지 실험할 수 있었다. 이는 건물과 풍경 전체에 원근감 효과와 풍부한 전망을 창출할 기회를 만들었다. 타일의 단점은 특정 상황에서 쉽게 깨진다는 점이나, 이미 의도적으로 깨진 채로 적용되면 이 점은 무관하다.

GLASS-Q1: Tell us about your favourite project that you used glass in or another architect's work - interior, facade, etc.

A: For the Albanian National Museum of Photography, Marubi, we have designed a curtain wall to cover the three facades of the central courtyard of the building that has helped to define its strong identity. Thanks to a specially designed structure, the partitions of the glazed facades follow a rhythmic pattern where the structure is

A: 알바니아 국립 사진 박물관 마루비의 경우, 우리는 건물의 강한 정체성을 정의하는 데 도움이 된 커튼월을 디자인했다. 이는 건물 중앙 안뜰의 세 정면을 덮는다. 특별히 설계된 구조 덕분에 구조가 내장되어 눈에 보이지 않고, 유리 파사드의 파티션은 리드미컬한 패턴을 따른다.

Albanian National Museum of Photography
©Christian Richters

embedded and remains hidden to the eye. In that way, the glazed facades seem to be free from structural requirements and their pattern can expressively be used to create a rich dialogue not only among the glazed facades themselves but also between them and the rest of the building.

The courtyard becomes a 'musical' architectural element, bathed by the changing natural light captured by the glazed facades.

At night, through the transparency of the glazed facades, lights from inside the building bathe the space of the courtyard and create an intimate place for relaxing and for enjoying the decaying light of the sunset.

GLASS-Q2: What are the strengths and weaknesses of glass?

A: As mentioned before, for us the potentials of using glass are enormous.
Glasses can be transparent or can present a wide range of translucency, they can be colored, can be printed, can create rich reflections or can be matt.
Nowadays it is possible for the designer to express very different architectonic

intentions by using different glass qualities, by combining glasses of different properties, by curving them or by creating contrast with other materials that will be reflected on them. **Glass is the most versatile material in our opinion.** With its properties, it translates poetical intentions into architecture in a unique way.

른 성질의 유리를 결합하거나, 곡선을 만들거나, 다른 재료와 대조를 만들어 매우 다른 건축적 의도를 표현할 수 있다. **우리 생각에 유리는 많은 재료 중 가장 다재다능한 재료이다.** 유리는 독특한 방식으로 시적 의도를 건축으로 번역한다.

WOOD-Q1: Tell us about your favourite project that you used wood in or another architect's work - interior, facade, etc.

A: At this moment our office is developing a museum on an existing platform on the waters of a lake in Shkodra, Albania. This museum is part of a bigger development, also developed by our office, in which the intervention area covers an extension of 3,7 km of the shore of Shkodra Lake and 5 km of a new pedestrian and bicycle path in the mountains that runs parallel to the shore. Both together, the renewed waterfront and the new mountain route create a circular pedestrian and bicycle route of 10 km that offers locals and visitors alike a wide variety of recreative, cultural and tourist facilities. These facilities

A: 현재 우리 사무실은 알바니아 슈코더르(Shkodra)에 있는 호수가에 현존하는 대지에 박물관을 디자인하는 중이다. 이 박물관은 우리 사무실이 맡은 더 큰 개발지 일부이다. 사이트는 슈코더르 호수 해안 3.7km와 산 중에 위치한 해안과 평행한 새로운 보행자 및 자전거 경로로 5km가 포함된다. 새로워진 해안가와 새로운 산악길은 지역 주민과 방문객 모두에게 다양한 레크리에이션, 문화 및 관광 시설을 제공하는 10km의 보행자 및 자전거용 원형 경로를 만든다. 이 시설은 22개의 전시학적 인터벤션으로 연결된 21개의 건축 및 조경 인터벤션으로 나누어

have been divided into 21 architectural and landscape interventions, which are connected to 22 museographic interventions. One of those architectural interventions is this museum made mainly of wood.

The selection of wood for the facades of the museum allows us to integrate it better in its natural environment. The facades are made with charred Western Red Cedar planks and wooden bars of untreated Siberian Larch on top of the planks. The charring process forms a black, sober appearance on the planks, and the untreated wooden bars on top of them create contrast in color and texture. This will create a vibrant facade in neutral colors that will change over time. The facade will reflect the passage of time and will get more integrated into the colors of the surroundings

져 있다. 이러한 건축 인터벤션 중 하나는 주로 나무로 만든 이 박물관이다.

박물관의 파사드를 위해 목재를 선택함으로써 자연과 더 잘 통합할 수 있다. 파사드는 까맣게 탄 서부 붉은 삼나무 판자와 널빤지 위에 처리되지 않은 시베리아 낙엽송 막대로 만들어진다. 탄화 과정은 널빤지에 검은색의 진지한 외관을 형성하고, 그 위의 무가공 된 목재는 색상과 질감의 대비를 만든다. 이것은 시간이 지남에 따라 변할 중립적인 색상의 활기찬 외관을 만들 것이다. 이 파사드는 시간의 흐름을 반영하고 주변 환경의 색상에 더욱 더 잘 통합될 것이다.

WOOD-Q2: What are the strengths and weaknesses of wood?

A: The use of wood with the FSC mark is eco-friendly and help us to minimize the carbon footprint of the building in its natural environment. Its color and

A: FSC 마크가 있는 목재를 사용하는 것은 친환경적이며 건물의 탄소 발자국을 최소화하는 데 도움이 된다. 자연의 법칙을 따라 시간이 지나면서 나

texture, which will change with time following a natural process, will remind visitors the nests of the birds in the area and its smell will awake poetic resonances of the woods that were present in the past in the mountains bordering the lake.

On the one hand, wood is a material that provides comfort and that is loaded with poetic meanings, capable of evoking nature and of provoking healing experiences thanks to its visual properties, its touch, and its profound smell.

On the other hand, wood, being an eco-friendly material, helps designers to create more sustainable environments, with respect for nature.

무의 색과 질감은 변할 것이며, 방문객에게 이 지역에 서식하는 새의 둥지를 상기시키고 나무의 향은 호수와 인접한 산에 과거에 존재했던 숲의 시적 공명을 깨울 것이다. 한편으로 나무는 편안함을 제공하고 시적 의미가 넘치며, 자연을 상기시키고, 시각적 특성, 촉감 및 깊은 냄새 덕분에 치유 경험을 자극할 수 있는 재료이다.

다른 편으로 친환경적인 재료인 목재는 디자이너가 자연에 대한 존중하에 보다 지속 가능한 환경을 조성하는 데 도움이 된다.

Museum in Shkodra

CEBRA

©Photopop

Who is ...?

CEBRA is a Danish architectural office founded in 2001 by the architects Mikkel Frost, Carsten Primdahl and Kolja Nielsen. In April 2017, architect MAA Mikkel Hallundbæk Schlesinger entered the group of partners.

Based in Aarhus in Denmark and in Abu Dhabi in the UAE, CEBRA employs a multidisciplinary international staff of 50 architects, constructing architects, urban planners and landscape architects, who all share a strong passion for architecture.

Q1: What is material to an architect (or to you)?

A: The materials do not usually come first for us at CEBRA. Louis Kahn supposedly asked bricks what they wanted to become, but for us, it's the other way around. **We develop a concept first. Then we start thinking about what the most suitable material is for that specific idea or form.** To us, materials are conceptual enhancement and realization tools.

A: 보통 우리 CEBRA에서는 재료를 먼저 고르지 않는다. 루이스 칸(Louis Kahn)은 벽돌에게 뭐가 되고 싶은지 물었지만, 우리는 그 반대이다. **우리는 먼저 디자인 컨셉을 개발한 다음 특정한 아이디어나 형태에 가장 적합한 재료가 무엇인지 생각하기 시작한다.** 우리에게 재료는 개념적 향상과 실현 도구이다.

Q2: Tell us about your favourite (or most often used) material and why.

A: We do not have favorite materials. That would be like having a favorite color which is a bit childish. Colors only make sense in relation with other colors. This is when they work well or not, and the materials are the same way. They do not make sense without a context. And we never chose the sites on which we build, so only when one knows the site and the project a certain material becomes good or bad. Doing white buildings all over the world as some sort of signature like for instance Siza seems strange to us – like a blindness to the surroundings.

A: 우리는 좋아하는 재료가 없다. 이는 가장 좋아하는 색과 비슷하게 약간 유치하다. 색은 다른 색과 비교하여만 의미가 있다. 그래야 서로 잘 어울리거나 그렇지 않을 때를 알 수 있으며 재료도 똑같다. 콘텍스트 없이는 말이 되지 않는다. 그리고 우리가 프로젝트를 지을 사이트를 고르는 일은 없으므로 사이트와 프로젝트를 알아야만 특정 재료가 좋거나 나쁘다고 할 수 있다. 예를 들어 시자(Siza)가 일종의 서명처럼 전 세계의 흰색 건물을 짓는 것은 우리에게 이상하게 보인다. 이는 마치 주변 환경이 보

It turns out though that we've constructed quite a few houses from wood. Applied in different ways depending on the circumstances, of course.

이지 않는 것 같다. 깨닫고 보니 우리는 나무로 꽤 많은 집을 지었다. 물론 상황에 따라 다른 방식으로 적용했다.

Q3: When do you decide the material during the design process and what is your criteria? (e.g. budget, client's preference, design concept, climate, etc.)

A: **We usually develop the concept and form before we think too much about materials.** But no two design processes are ever the same and sometimes materials are defined by the client or planning regulations. In those situations, the material is a prime concern from day one. And I would be lying if I said that economy plays no part in our decisions. In fact, we've built a career on getting a lot out of small budgets – bang for the buck we call it!

A: **우리는 대개 재료를 너무 깊이 생각하기 전에 개념과 형태를 개발한다.** 그러나 모든 설계 과정이 서로 다르고, 때로는 의뢰인 또는 계획 규정에 의해 재료가 결정되는 경우가 있다. 이런 상황에서는 그 재료가 첫날부터 가장 중요한 관심사가 된다. 그리고 경제적인 면이 우리의 결정에 아무런 영향을 미치지 않는다고 하면 거짓말이다. 사실 우리는 적은 예산을 최대한 활용하는 것으로 경력을 쌓았다. 우리는 본전을 뽑는다고 부른다!

Q4: What are some architectural projects that inspired you regarding brick, tile, wood and/or glass? And why?

A: We are inspired all the time by all sorts of things, so it is hard to pick out one

A: 우리는 항상 모든 것에서 영감을 받기 때문에 특정한 예를 들기 어렵다. 맨 처

particular example. I guess the first that comes to mind is Herzog & De Meuron's Dominos Winery where they built stone facades in a new way. That project seems to bridge the gap between tradition and avant garde as it reinterprets old mason walls. I quite like that duality. We've also been rather taken by the amazing brick facades Gramazio & Kohler do with robots. That too is a new way of using an ancient material.

Q5: Tell us about the materials you are interested in or want to use in your projects right now.

A: We are building a huge project in Abu Dhabi right now and most of it is cast in concrete. That too is a great material when used the right way. We've developed a warm sandy mix and we combine it with natural stone and areas of sand in matching colors.
We are also exploring rammed earth walls and ceramic tiles right now but honestly, we tend to use a rather small palette of materials in our office. And most of these may look completely different depending on how they are treated or applied. A brick for instance can have a million

different colors and surfaces and might even be glazed. Then you can play around with the mortar between the bricks not to mention the bond. We only need a few old trusted friends to get the job done. And we like knowing how materials perform over time. How they hold up, how they age and how they effect the environment. If one sticks with the basics all of these factors should be relatively well known.

BRICK-Q1: Tell us about your favorite project that you used brick in or another architect's work - interior, facade, etc.

A: We've used bricks in many of our projects. In Denmark, most houses are built from this material so it's deeply rooted in our culture. I think we managed to reinterpret the material and building system in our office building for the tech company Elbek & Vejrup. Here the bricks are applied like digital pixels and have multiple bright colors. It's almost like a Minecraft aesthetic.

유약 처리를 할 수도 있다. 그런 다음 접착제는 물론 벽돌 사이의 모르타르로도 여러 가지 실험해 볼 수 있다. 일을 끝내기 위해 신뢰하는 오랜 친구 몇 명만 필요하다. 그리고 우리는 재료가 시간이 지남에 따라 어떻게 되는지 아는 것을 좋아한다. 재료가 어떻게 버티고, 나이를 어떻게 먹고, 환경에 어떻게 영향을 미치는지 알고 싶다. 만약 기본을 고수한다면, 이 모든 요소를 비교적 잘 알 수 있다.

A: 우리는 많은 프로젝트에서 벽돌을 사용했다. 덴마크에서는 주택 대부분을 벽돌로 지으며 우리 문화에 깊이 뿌리를 둔다. 우리는 테크 회사인 엘벡 앤드 베이럽 (Elbek & Veirup)의 사무실 건물에서 벽돌과 그 건축 시스템을 재해석할 수 있었다. 여기서 벽돌을 디지털 픽셀처럼 적용했으며 여러 가지 밝은 색상을 썼다. 거의 마인크래프트의 미학 같다.

Elbek & Veirup

152 Brick, Brick! What do you want to be?

BRICK-Q2: What are the strengths and weaknesses of brick?

A: It doesn't quite make sense to me to talk about weaknesses when it comes to materials. **Different materials simply have different properties – like musical instruments.** Drums are amazing for rhythm, but if you need to play a melody you might need something like a guitar. So, materials are all great, but they need to be applied properly and used in the right situation. For bricks specifically, I'm often faced with economical challenges. If I manage to find a stone that is just right in terms of color and texture, it is often over budget and I have to settle for something else. Making beautiful details with brick also cost labor hours and is, unfortunately, a rare possibility.

A: 재료를 논할 때 단점을 이야기하는 것은 말이 되지 않는다. **여러 재료는 악기와 같이 다른 속성을 가지고 있을 뿐이다.** 드럼은 리듬에 아주 적합하지만, 멜로디를 연주해야 한다면 기타와 같은 것이 필요할 수 있다. 따라서 재료는 모두 훌륭하지만 올바르게 적용되고 적절한 상황에서 사용되어야 한다. 특히 벽돌을 쓸 때 나는 종종 경제적 어려움에 직면한다. 색깔과 질감 면에서 딱 맞는 벽돌을 찾으면 종종 예산을 넘어서 다른 제품으로 만족해야 한다. 벽돌로 아름다운 디테일을 만드는 것은 또한 노동 시간이 많이 소요되며 불행히도 달성할 가능성이 매우 작다.

TILE-Q1: Tell us about your favourite project that you used tile in or another architect's work - interior, facade, etc.

A: We've used tiles graphically in a few of our projects – like digital pixels. At the Elbek & Vejrup project, we also made black space invaders on the bathroom

A: 우리는 몇 가지 프로젝트에서 타일을 디지털 픽셀처럼 그래픽으로 사용했다. 엘벡 앤드 베이럽 프로젝트에서 우리는 욕실 바닥에 흑색의 침

CEBRA 153

floors and in the Experimentarium project we created optical illusions with tiles. It´s a cheap but visually very powerful way to twist spaces and beef up dull spaces.

TILE-Q2: What are the strengths and weaknesses of tile?

A: In cold climates like here in Denmark applying tiles to an exterior façade can prove technically difficult. In the very few places I know where architects have stuck their neck out the tiles simply come clattering down.

GLASS-Q1: Tell us about your favourite project that you used glass in or another architect's work - interior, facade, etc.

A: Apart from the glass, you´ll find in the windows of our projects, we do not use the material a lot.
I remember visiting Maison de Verre in Paris as a student some 25 years ago. It made a great impression on me. The house is designed by Pierre Chareau in 1928 and it was decades ahead of time in terms of transparency effects, details and

aesthetics. I believe it is open to the public today and it is certainly worth a visit.

측면에서 수십 년 앞서있었다. 오늘날에도 대중에게 공개된 것 같고 확실히 방문할 가치가 있다고 생각한다.

Maison de verre ©Subrealistsandu

GLASS-Q2: What are the strengths and weaknesses of glass?

A: The building codes in Denmark are rather strict when it comes to energy efficiency. So, when we design big glazed window areas, we struggle with heat loss in the winter and overheating in the summer. Glass as a cladding material is quite splendid though. We used it for

A: 덴마크의 건축법은 에너지 효율에 관해서 다소 엄격하다. 그래서 우리는 큰 유리창을 설계할 때 겨울에는 열 손실로, 여름에는 과열로 고생한다. 하지만 클래딩 재료로서 유리는 상당히 훌륭하다. 우리는 그룬드포스 전문대(Grundfos Kollegiet)의 거대

our great mirror space in The Grundfos Kollegiet. It's funny though, glass is probably the only material that an architect cannot totally avoid. It would be like cooking without salt.

한 거울 공간에 유리를 사용했다. 재밌는 점은 건축가가 완전히 피할 수 없는 유일한 재료는 유리일 것이라는 점이다. 아마 소금 없이 요리하는 것과 같을 것이다.

WOOD-Q1: Tell us about your favourite project that you used wood in or another architect's work - interior, facade, etc.

A: There are many fantastic wooden buildings in the world – both new and old. Lately, I have been particularly fascinated by the wooden structures designed by Kengo Kuma. The structural qualities and the filigree patterns he makes are just amazing.

I am also quite happy with our own LIF project. It's small and bit overlooked but as it happens it became "the mother" of many bigger CEBRA projects such as the smart school in Irkutsk. Here structure and cladding become one in a vertical symbiosis that is quite successful.

A: 세계에 예전 것이든 최근 것이든 멋진 목조 건물이 많이 있다. 최근에 나는 특히 쿠마 켄고가 디자인한 목조 구조물에 매료되었다. 그가 만드는 구조적 특성과 줄 패턴은 그저 놀라울 뿐이다. 나는 우리의 LIF 프로젝트에도 매우 만족한다. 이 프로젝트는 작고 약간 간과되었지만, 실제로 러시아 이르쿠츠크(Irkutsk)의 스마트 스쿨과 같이 더 큰 CEBRA 프로젝트 다수의 "어머니"이다. 여기서 구조와 클래딩은 수직 공생으로 하나가 되었고 이는 꽤 성공적이었다.

WOOD-Q2: What are the strengths and weaknesses of wood?

A: Fire regulations have been a challenge for many years – if one builds many floors constructed by wood. Many clients also complain about maintenance saying that wood needs constant care, but I disagree with that. **In many ways wood is a fantastic material. You can use it for almost anything.** It's cheap and easy to work with and in terms of structure and statics it performs really well. The funny thing is that it just grows out there in the forest. You simply need to cut it down and it's almost ready for use – no baking, no mixing or reinforcing. If you need more, you simply plant it and wait. In the meantime, it's a pleasure to watch and a home for birds. It even produces oxygen for the planet!

Wood is not ideal in all situations – as stated above, no material is – but it comes pretty close. I've just had my whole yard covered with hard wood. It's great to walk on it barefoot and it heats up nicely in the sunlight.

A: 나무로 많은 층을 지으려면 수년 동안 소방법이 난제였다. 많은 의뢰인이 목재에 지속적인 관리가 필요하다며 유지 보수에 대해 불평하지만, 나는 그에 동의하지 않는다. **여러 면에서 나무는 환상적인 재료이다. 거의 모든 것에 사용할 수 있다.** 나무는 저렴하고, 작업하기 쉽고, 구조와 공전적인 면에서 매우 좋다. 재미있는 것은 나무는 숲에서 그냥 자라는 것뿐이라는 점이다. 그냥 잘라내면 사용할 준비가 거의 되어있다. 구울 필요도, 섞을 필요도, 강화할 필요도 없다. 더 필요하면 그냥 심고 기다리기만 하면 된다. 자라는 동안 나무를 보는 것은 즐거운 일이고 새를 위한 집이 된다. 심지어 지구를 위해 산소까지 생산한다! 위에서 언급했듯이 모든 상황에 이상적인 재료는 없고 나무 역시 그렇지 않지만 완벽함에 꽤 가깝다. 나는 얼마 전 내 마당 전체를 원목 마루로 깔았다. 마루는 맨발로 걷기 아주 좋고 햇볕에 적당히 따뜻해진다.

Davide Macullo Architects

Who is ...?

Davide Macullo (b. Giornico, CH, 1965) lives and works in Lugano, Switzerland.

Studied art, architecture and interior design. For 20 years (1990-2010) he was project architect in the atelier of Mario Botta with responsibility for over 200 international projects worldwide. He opened his own atelier in 2000.

Q1: What is material to an architect (or to you)?

A: Materials raise feelings related to one's experiences and sensibility, they also express techniques and innovations discovered by Man, and describe historical moments. **As in the architecture of our ancestors, the use of materials should represent and symbolize cultural values.**

A: 재료는 경험 및 감수성과 관련된 감정을 높이고, 인간이 발견한 기술과 혁신을 표현하며, 역사적인 순간을 묘사한다. **조상들의 건축에서와 마찬가지로, 재료는 문화적 가치를 대표하고 상징해야 한다.**

Q2: Tell us about your favourite (or most often used) material and why.

A: **All natural materials that relate to the context in which we use them are our favorites.** The ideal is when the material used belongs to the region we build.

A: **우리는 프로젝트의 콘텍스트와 관련된 모든 천연 재료를 좋아한다.** 우리의 이상은 사용하는 재료가 건물을 짓는 지역에서 나올 때이다.

Swisshouse XXXII Interiorroof ©Alexandre Zweiger

Davide Macullo Architects

Q3: When do you decide the material during the design process and what is your criteria? (e.g. budget, client's preference, design concept, climate, etc.)

A: **We prefer indigenous materials.** Their technical characteristics related to their use and budget is fundamental. The final choice of the materials is suggested by the project itself. First we plan the spaces in relation to the psychology of the people and the context; the choice of materials follows.

A: 우리는 **지역 고유의 재료를 선호**하며 건물의 용도에 관련된 기술적 특성 및 예산이 필수적인 기준이다. 재료의 최종 선택은 프로젝트 자체가 시사한다. 우리는 먼저 사람들의 심리와 콘텍스트에 기반하여 공간을 계획하고 그 후 재료 선택이 뒤따른다.

Q4: What are some architectural projects that inspired you regarding brick, tile, wood and/or glass? And why?

A: The Roman church (year 1176) in Giornico, where I was born, has influenced my perception of modernity. This church, completely built in stone, is an example how a building with over 1000 years of history can be still completely contemporary. It refers to a sustainable attitude, it integrates art with architecture, it is beautiful, meant for a poetic purpose, it is full of symbolism and it is a jewel of technique, including a great acoustics. It is a masterpiece. Also

A: 내가 태어난 스위스 조르니코(Giornico)에 지어진 로마 교회(1176년)는 근대성에 대한 나의 인식에 영향을 미쳤다. 완전히 돌로 지어진 이 교회는 1000년이 넘는 역사를 가진 건물이 어떻게 여전히 완벽하게 현대적일 수 있는지 보여준다. 이는 지속 가능한 태도를 뜻하고, 예술과 건축을 통합하고, 아름답고, 시적 목적을 위한 것이며, 상징성으로 가득하며, 훌륭한 음향을 포함한 기술의 보석이다.

The Roman church (year 1176) in Giornico

the pathway to the Temple of Heaven in Beijing that has been built about 17th century; a kind of bridge over the land of about 3 meters high and one kilometer long, that gives man the feeling of flying, only by using the intelligence and care in managing and manipulating the senses of the visitors. The tomb of Agamennon in Greece in Mycenae is another example of a simple use of materials that can change your way of understanding life. It is like you give a glimpse of an eye of your own tomb. It is so contemporary in describing through the senses the path between life and death.

Wood has so many beautiful and fascinating examples throughout the history that it is so difficult to select an example. The space where I used to hide

이 교회는 걸작이다. 또한 약 17세기에 지어진 베이징의 천국의 성전으로 가는 길도 굉장하다. 높이가 약 3m, 길이가 1km인 일종의 땅 위의 다리로, 방문객의 감각을 관리하고 조작하는데 지능과 관심을 기울인 것만으로 마치 날아가는 느낌을 준다. 그리스 미케네(Mycenae)에 있는 아가멤논의 무덤은 삶을 이해하는 방식을 바꿀만한 간단한 재료 사용법의 또 다른 예이다. 마치 자신의 무덤을 언뜻 보는 것 같고 감각을 통해 삶과 죽음 사이의 길을 묘사하는 방식이 매우 현대적이다. 나무는 역사를 통틀어 아름답고 매혹적인 예가 너무 많아서 하나만 선택하기가 어렵다. 유치원 때 내가 숨곤 했던 공간은 세 개의 노송나무 사이 공간이었다. 거대한 나무

Gate Temple of Heaven in Beijing ©Hermann Luyken

as a child in kindergarten was a space between three cypresses. Three immense nails of wood. This was my first strong experience with sensing a space. It was like I could be projected from my little village to anywhere in the world. This was my way to follow my dreams and I understood that three lines define a space. Glass is a conceptual material, Dan Graham (the artist) has taught me the power of this material with his wonderful research of putting man in relation to transparencies and reflections.

손톱 세 개 같았는데 이는 내가 최초로 공간을 감지한 강력한 경험이었다. 마치 내가 우리 작은 마을에서 세계 어디로든 투사될 수 있을 것 같았다. 이것이 내가 꿈을 좇는 방법이었고 나는 그 세 줄이 공간을 정의한다는 것을 이해했다.
유리는 개념적 소재이며, 댄 그레이엄(예술가)은 투명성과 반사와 관련하여 사람을 두는 훌륭한 연구로 이 재료의 힘을 가르쳐주었다.

Pavillon_Dan Graham ©Wikipedia Commons

Q5: Tell us about the materials you are interested in or want to use in your projects right now.

A: Any materials that have a positive influence on Man interest me.

A: 인간에게 긍정적인 영향을 미치는 재료는 모두 흥미롭다.

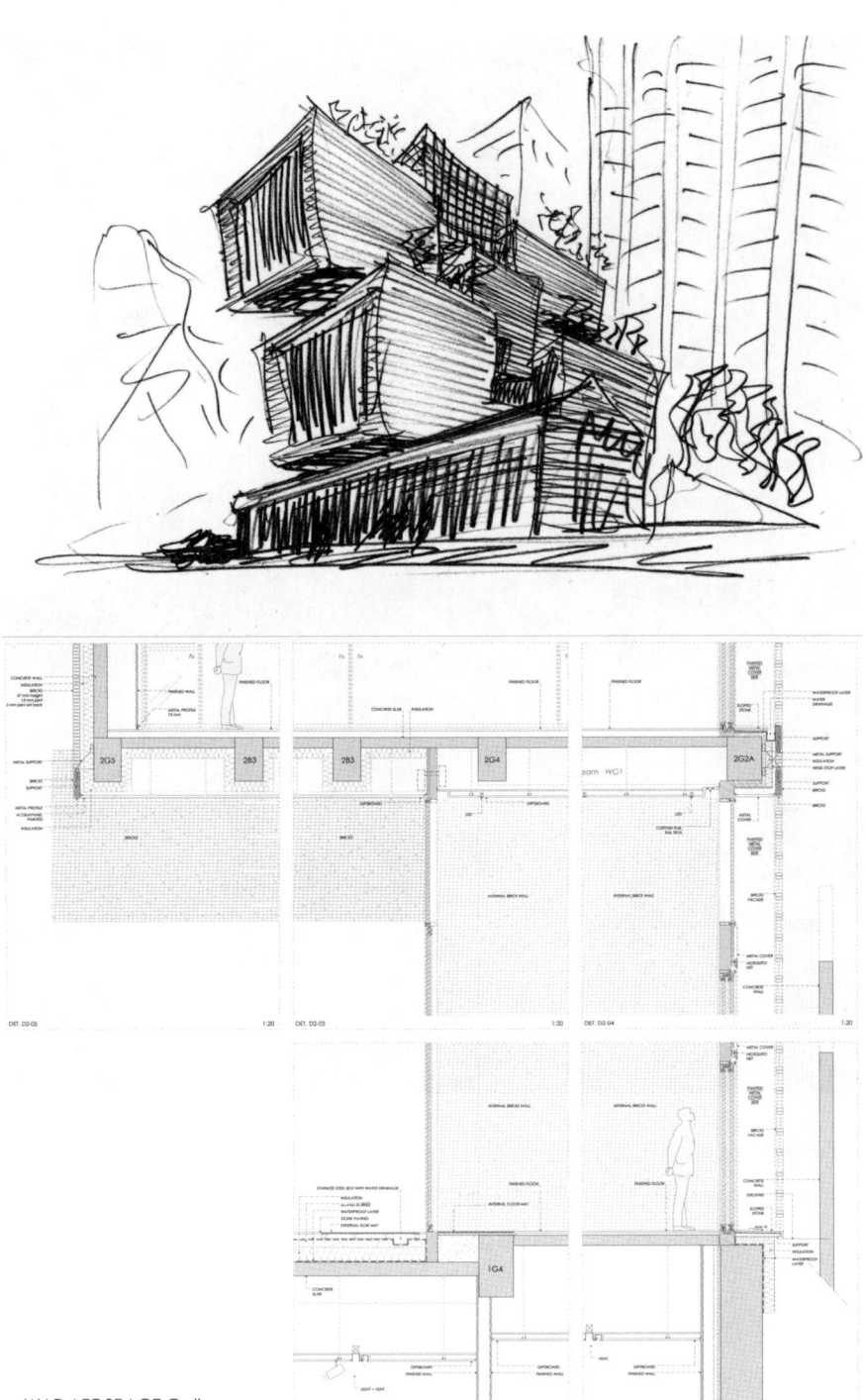

WAP ART SPACE Gallery

164 Brick, Brick! What do you want to be?

BRICK-Q1: Tell us about your favorite project that you used brick in or another architect's work - interior, facade, etc.

A: The WAP ART SPACE gallery that we recently built in Gangnam Seoul is an interesting use of the bricks as we intend it. The bricks are like a silk cloth you wear and give the feeling of keeping the body warm or cool. The impact of bricks on the elevations of a building is connected to the reduction of scale: when I see the building from far it has a homogeneous appearance, the closer I come to the building the more I can distinguish the characteristics of the cladding material. This holds my attention during the different phases of the approach and reduces the scale from urban to domestic, from public to private. The more I get closer, the more I can nurture my feelings. **Bricks have the extraordinary potential in expressing the human work. Is like seeing a fabric woven by hand instead of by machine. It is alive and expresses a care for Man.**

A: 최근 서울 강남에 지은 WAP ART SPACE 갤러리는 우리의 의도대로 벽돌을 흥미롭게 사용할 수 있었다. 벽돌은 몸에 걸치는 실크 천과 같고 몸을 따뜻하게 혹은 시원하게 유지하는 느낌을 준다. 건물 입면에 벽돌이 미치는 영향은 축척의 축소와 연관된다. 멀리서 건물을 보면 균질한 외관이지만, 건물에 가까이 올수록 클래딩 재료의 특성을 더 구별할 수 있다. 이는 다른 단계의 접근을 거치는 동안 계속해서 나의 관심을 끌고, 도시에서 가정, 공공에서 개인으로 축척을 축소한다. 가까이 다가갈수록 내 감정을 더 잘 느낄 수 있다. 벽돌은 인간의 작품을 표현할 수 있는 특별한 잠재력을 가지고 있다. 기계가 아닌 손으로 짠 천을 보는 것과 같다. 벽돌은 살아 있고 인간에 대한 보호를 잘 표현한다.

BRICK-Q2: What are the strengths and weaknesses of brick?

Davide Macullo Architects

A: In our practice we build up an awareness of emphasizing the strength of the materials or using the materials for their adaptability to our goals. So we don't care about weakness.

A: 우리 사무소는 재료의 장점을 강조하거나 우리의 목적에 맞는 적응력을 가진 재료를 쓰는 것에 대한 인식을 쌓기 때문에 재료의 단점에 대해서는 신경 쓰지 않는다.

TILE-Q1: Tell us about your favourite project that you used tile in or another architect's work - interior, facade, etc.

A: The only tiles we consider as quality product are the hand produced ones. I think that besides the beauty of the historical buildings around the world that have seen wonderful tile works, in the recent history Alvaro Siza has been able to use the tiles in a simple but extraordinary way. For example in the church of San Marco de Canavezes in Portugal.

A: 우리가 고품질 제품으로 간주하는 유일한 타일은 수제 타일이다. 전 세계 역사적인 건물의 아름다움에서 볼 수 있었던 멋진 타일 작품 외에도 최근 역사에서 알바로 시자(Alvaro Siza)가 단순하지만 특별한 방식으로 타일을 사용했다고 생각한다. 예를 들면 포르투갈의 산 마르코 데 카나베즈(San Marco de Canavezes) 교회가 있다.

San Marco de Canavezes
©Manuel Anastácio

TILE-Q2: What are the strengths and weaknesses of tile?

A: The handmade tiles are simply a pleasure for the senses. The industry produced tiles are aggressive and don't belong to the human space because it is a product that express an industrial process without soul and this is psychologically negative for the environment of people to live within.

A: 수제 타일은 단순히 감각을 위한 즐거움이다. 대량 생산된 타일은 공격적이며 영혼이 없는 산업 과정을 표현하는 제품이기 때문에 인간의 공간에 속하지 않는다. 이는 사람들이 살 환경 내부에 쓰기에는 심리적으로 부정적이다.

GLASS-Q1: Tell us about your favourite project that you used glass in or another architect's work - interior, facade, etc.

A: Glass is a natural material. Glass represents first of all a transparence between inside and outside, secondly, a potential of using it for its reflection characteristics. An interesting recent project in glass is the work of Jean Nouvel for the entrance wall of the Foundation Cartier in Paris for its dimension and relation between two spaces that are both outdoor spaces but in two different conditions; street-garden. In our house in Muzzano we used the same principle in a reduced intimate space where the glass has

A: 유리는 천연 물질이다. 가장 첫째로 유리는 내부와 외부 사이의 투명성을 나타내며, 두 번째로 반사 특성을 위해 사용할 가능성이 있다. 최근 흥미로운 유리 프로젝트는 장 누벨(Jean Nouvel)이 파리 카르띠에 재단의 입구 벽에 한 작업이다. 두 가지 조건의 두 야외 공간 사이의 차원과 관계가 흥미롭다. 마치 거리 정원 같다. 무자노(Mazzano)에 있는 우리 집에 좀 더 작고 친밀한 공간에서 같은 원리를 사용했고 여기서 유리는 반사적 특성을 위해 쓰였다. 둘 다 예술가 댄 그

Davide Macullo Architects 167

been used for its reflecting characteristics. Both are inspired by the work of the artist Dan Graham.

GLASS-Q2: What are the strengths and weaknesses of glass?

A: Glass is the material of wonder: it can be the most solid or the most inexistent material, the most challenging and the most dangerous material. Too much reflection and transparence is against human nature. The extensive use of glass in building industry has brought the rise of a series of uncomfortable situations where people have increased their level of stress. I refer to situations such as glass buildings in metropolitan areas, for example London office spaces suffer from this extended use of glass for the working habitat of humans.

WOOD-Q1: Tell us about your favourite project that you used wood in or another architect's work - interior, facade, etc.

A: We have just built a house "Swisshouse XXXII" in the Swiss mountains together

Swiss House IV (House in Muzzano) ©Pino Musi

with the artist Daniel Buren. The wood structure recalls the verticality of the forests and we used local wood. The archetype of the house has been developed one step further and the collaboration with the artists makes this building a sculpture and a house at same time. It is a sculpture to live within. For hundreds of years we have discussed about integration between art and architecture and this is one of the rare examples where the two arts really cooperate. Without the art the building doesn't exist because the art is a structural part of the building. As we have the walls designed in collaboration with Daniel Buren, the beams of the roof are designed in collaboration with another conceptual artist from the region, Miki Tallone.

위스 산맥에 "스위스 하우스 XXXII"라는 집을 지었다. 나무 구조는 숲의 수직성을 상기시키며, 그 지역의 목재를 사용했다. 집의 원형을 한 걸음 더 발전시켰고 예술가와의 협력으로 이 건물은 조각품인 동시에 집이 되었다. 안에서 살 수 있는 조각인 것이다. 수백 년 동안 예술과 건축의 통합이 논의되어 왔고 이 건물은 두 예술이 실제로 협력하는 드문 예 중 하나이다. 이 건물은 예술이 건물의 구조적 부분이기 때문에 예술이 없으면 건물은 존재하지 않는다. 이미 다니엘 뷰런과 공동으로 설계된 벽이 있었기에 지붕의 보는 그 지역의 다른 콘셉트 예술가인 미키 탈론(Miki Tallone)과 공동으로 설계했다.

Construction process of Swisshouse XXXII ©Alexandre Zweiger

WOOD-Q2: What are the strengths and weaknesses of wood?

A: Wood is one of the most beautiful materials because it is the closest to nature and Man in its use.

A: 나무는 그 용도에 있어서 자연과 인간에 가장 가깝기 때문에 가장 아름다운 재료 중 하나이다.

Swisshouse XXXII ©Alexandre Zweiger

Davide Macullo Architects

Donner Sorcinelli Architecture

Who is ...?

Donner Sorcinelli Architecture is an international architectural design office based in Italy.

Founded by architects Luca Donner and Francesca Sorcinelli, the firm pays particular attention to the theme of sustainable and affordable architecture in all its variants, based on experimentation and research in various fields like Architecture, Urban Design, Interior and Product Design.

Q1: What is material to an architect (or to you)?

A: **Any material is not only a technological expression of the specific design but rather its real poetics.**

A: 모든 재료는 특정 디자인의 기술적 표현일뿐만 아니라 진정한 시적 표현이다.

Q2: Tell us about your favourite (or most often used) material and why.

A: Wood is not the only one we use to use but what we love more because it is natural, recyclable, versatile in its applications.

A: 우리는 나무만 쓰는 것은 아니지만, 자연스럽고 재활용이 가능하며 다양한 용도로 쓸 수 있기 때문에 아주 좋아한다.

Q3: When do you decide the material during the design process and what is your criteria? (e.g. budget, client's preference, design concept, climate, etc.)

A: Availability of the material on site is a key factor to be considered. This aspect is related to a sustainable approach that we keep in mind in any design process. Budget is always important in order to provide an affordable solution.

A: 현장에서 재료의 가용성은 꼭 고려해야 할 핵심 요소이다. 이 점은 어떤 설계 과정에서든 염두에 두는 지속가능한 디자인 방식과 연관이 있다. 예산은 실현 가능한 해결책을 내기 위해 항상 중요하다. 심리와 콘텍스트에 기반하여 공간을 계획하고 그 후 재료 선택이 뒤따른다.

Q4: What are some architectural projects that inspired you regarding brick, tile, wood and/or glass? And why?

A: There are many projects which are a reference for us in relationship to those materials but the main ones are:
Brick and Tiles – Ningbo Museum by Wang Shu
Glass – Nelson-Atkins Museum of Art by Steven Holl

A: 재료와 관련하여 우리에게 참고가 된 프로젝트는 많지만 주요 프로젝트는 다음과 같다.
벽돌과 타일 –
왕수(王樹)의 닝보 박물관
유리 –
스티븐 홀(Steven Holl)의 넬슨-앳킨스 미술관 (Nelson-Atkins Museum of Art)

Ningbo Museum ©Siyuwj

Q5: Tell us about the materials you are interested in or want to use in your projects right now.

A: We are currently developing some projects in hot and humid regions where bamboo is available on site.

A: 우리는 지금 현장에서 대나무를 구할 수 있는 덥고 습한 지역에서 몇 가지 프로젝트를 진행 중이다.

Nelson-Atkins Museum of Art by Steven Holl ©Charvex

BRICK-Q1: Tell us about your favorite project that you used brick in or another architect's work - interior, facade, etc.

A: Two types of Brick Masonry have been used for the New National Museum of Afghanistan. Stone bricks have been used for the exterior while clay bricks for the interior layer of the building envelope.
The reason was to exploit only local materials in order to reduce the CO_2 footprint of the building as well as to use local workforce.

A: 아프가니스탄의 새로운 국립 박물관에 벽돌 쌓기의 두 가지 유형을 사용했다. 건물 외부에 조적 벽돌을, 건물 외피의 내부 층에는 점토 벽돌을 사용했다. 지역 재료만을 활용한 이유는 건물의 CO_2 footprint을 줄이고 지역 인력을 활용하기 위해서였다.

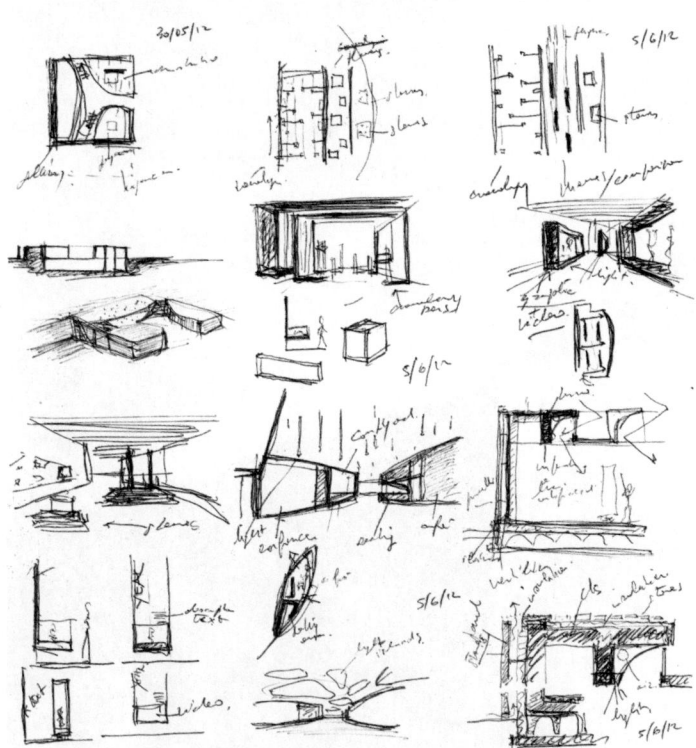

New National Museum of Afghanistan

BRICK-Q2: What are the strengths and weaknesses of brick?

A: Natural properties are pros as well as the intense work required is cons.

A: 자연스러운 특성이 장점이며, 손이 많이 가는 점이 단점이다.

TILE-Q1: Tell us about your favourite project that you used tile in or another architect's work - interior, facade, etc.

A: CD House roof has been built with large size flat dark grey ceramic tiles not so usual for the local environment which is usually characterized by light brown Spanish colonial ones.

A: CD 하우스 지붕은 일반적으로 밝은 갈색의 스페인 식민지풍 지붕이 특징인 그 지역에서는 평범하지 않은, 크기가 크고 평평한 진한 회색 세라믹 타일로 지었다.

TILE-Q2: What are the strengths and weaknesses of tile?

A: They are easy to assembly on the roof but easily affected by the local weather conditions.

A: 지붕에서 조립하기 쉽지만, 기상 조건에 쉽게 영향을 받는다.

GLASS-Q1: Tell us about your favourite project that you used glass in or another architect's work - interior, facade, etc.

CD House Roof Tiles in Progress

A: The use of glass always represents an exciting challenge to be played in between transparencies and reflections. One of the project where the combined use of glass and other materials has been developed according to these aspects it was the design of a Primary School in Carbonera (Italy). In this case, the visual connection between the common spaces and classrooms with the landscape has been the main driver of the whole design process.

A: 유리의 사용은 항상 투명도와 반사 사이에 생기는 흥미진진한 문제를 나타낸다. 이러한 측면에서 유리를 다른 재료와 결합해 사용했던 프로젝트 중 하나는 이탈리아 베네토 카르보네라 (Carbonera)의 초등학교 디자인이었다. 이 경우 풍경과 공용 공간 및 교실 간의 시각적 연결이 전체 설계 과정의 주요 동인이었다.

Primary School in Carbonera

GLASS-Q2: What are the strengths and weaknesses of glass?

A: There are no strengths or weaknesses but poetics' potentials only.

A: 유리에 장점이나 단점은 없다. 오직 시적 잠재력뿐이다.

WOOD-Q1: Tell us about your favourite project that you used wood in or another architect's work - interior, facade, etc.

A: Cross laminated timber panels have been used for CD House's structure in order to achieve high thermal insulation and environmental friendly design. Time and costs of construction have been reduced and kept under control.

A: 높은 단열성 및 환경친화적인 디자인을 위해 CD House의 구조에 직교적층목재 패널을 사용했다. 건설 시간과 비용을 감소하고 잘 관리할 수 있었다.

WOOD-Q2: What are the strengths and weaknesses of wood?

A: **Wood represents the sum of all qualities we use to require to a material.** It is strong enough to carry loads, it is natural, recyclable as well as warm by a visual and tactile point of view.

A: 나무는 우리가 재료에 요구하는 모든 자질의 합이다. 하중을 운반할 만큼 강하며, 시각 및 촉각적인 면에서 자연스럽고, 재활용할 수 있으며 따뜻하다.

CD House Under Construction

180 Brick, Brick! What do you want to be?

Katsutoshi Sasaki +Associates

Who is ...?
Office Information
4-61-3 Tanaka-cho
Toyota-shi Aichi
471-0845 Japan
Phone:+81 565 29 1521
sasaki@sasaki-as.com

1976 Born in Toyota-shi Aichi,Japan
1999 Guraduated Kindai University
2008 Established Katsutoshi Sasaki + Associates

Q1: What is material to an architect
    (or to you)?

A: A hearing about natural and human. and way of their connecting.

A: 자연과 인간에 대한 청문회 이자 이를 모두 연결하는 방식.

Q2: Tell us about your favourite
    (or most often used) material and why.

A: Wood. because one of nature.

A: 나무를 좋아한다. 자연의 일부니까.

Kinkaku-ji.jp © Martin Falbisoner

Q3: When do you decide the material during the design process and what is your criteria? (e.g. budget, client's preference, design concept, climate, etc.)

A: budget and design concept.

A: 예산 및 디자인 컨셉.

Q4: What are some architectural projects that inspired you regarding brick, tile, wood and/or glass? And why?

A: I was effected many architecture in the world.
My favorite architecture have innocence.

A: 나는 세상의 많은 건축에서 영향을 받았다.
내가 가장 좋아하는 건축은 순수함을 지녔다.

Q5: Tell us about the materials you are interested in or want to use in your projects right now.

A: Natural light.

A: 자연광.

Natural light in Tnoie

BRICK-Q1: Tell us about your favorite project that you used brick in or another architect's work - interior, facade, etc.

A: I don't have favorite it, but I like half screen architecture by the brick.

A: 가장 좋아하는 작품은 없지만, 벽돌로 만든 부분적 칸막이를 좋아한다.

BRICK-Q2: What are the strengths and weaknesses of brick?

A: Weakness point is high cost in japan. Strength point is deep texture.

A: 단점은 일본에서는 비싸다는 점이고, 장점은 깊은 질감이다.

TILE-Q1: Tell us about your favourite project that you used tile in or another architect's work - interior, facade, etc.

A: Our project Phiaro used tile made by soil of the land.

A: 우리 프로젝트 피아로 (Phiaro)는 현지의 토양으로 만든 타일을 사용했다.

TILE-Q2: What are the strengths and weaknesses of tile?

A: Weakness point is high cost in japan. Strength point is deep texture.

A: 단점은 일본에서는 비싸다는 점이고, 장점은 깊은 질감이다.

GLASS-Q1: Tell us about your favourite project that you used glass in or another architect's work - interior, facade, etc.

A: National Gallery in Germany by Mies van der Rohe.

A: 미즈 반 데어 로에 (Mies van der Rohe)의 독일 국립 미술관.

GLASS-Q2: What are the strengths and weaknesses of glass?

A: Weakness point is Heat insulating performance. There are a lot of good points.

A: 단점은 단열 성능이다. 장점은 많이 있다.

National Gallery in Germany ©Jean-Pierre Dalbéra

WOOD-Q1: Tell us about your favourite project that you used wood in or another architect's work - interior, facade, etc.

A: Japanese classical temple.

A: 전통적인 일본 절.

WOOD-Q2: What are the strengths and weaknesses of wood?

A: Weakness point is soft. Strength point is soft.

A: 단점도 장점도 모두 너무 부드럽다는 점이다.

Under Construction

Keiichi Hayashi Architect

Who is …?
 1967 Born in Osaka
 1991 Graduated from Metal Engineering, Kansai University
 1993 Graduated from Architecture, Kansai University
 1997 Established Keiichi Hayashi Architect

Design Philosophy It is important for me to make architecture using basic materials and uncomplicated construction methods. I try to create a system that is based on pure architecture but becomes complex when people use it.

Q1: What is material to an architect (or to you)?

A: **It gives character of physical environment into a space.**

A: 재료는 공간에 물질적 환경의 특징을 부여한다.

Q2: Tell us about your favourite (or most often used) material and why.

A: Corrugated sheet. It is very thin sheet material, which is only zero point a few mm steel plate change into strong material that holds several meters without bending with its simple deform processing. It is also surprisingly lightweight. This material provides high versatility since its wave size is standardized. It is also waterproof, easily processed and low-cost material. Many evidences listed above and being used mostly in regions of poverty prove that corrugated sheet is a fantastic

A: 파형판. 매우 얇은 시트 재료이며 단순한 변형 처리로 장스팬(span) 구부러지지 않고 유지하는 강한 재료로 겨우 몇 밀리미터 두께의 강판이다. 또한 놀랍도록 가볍다. 이 재료는 파형의 크기가 표준화되어 있기 때문에 용도가 다양하다. 방수가 되고 쉽게 가공되며 저렴한 재료이다. 위에 열거된 많은 이유와 빈곤 지역에서 주로 사용되는 점은 파형판이 환상적인 건축 자재임을 증명한다.

building material.

Q3: When do you decide the material during the design process and what is your criteria? (e.g. budget, client's preference, design concept, climate, etc.)

A: It will be naturally determined in the process of designing and it depends on the project. It is sometime determined by a first impression of the site. It is all for performance of material.

A: 설계 과정에서 자연스럽게 결정되며 프로젝트에 따라 다르다. 어떨 때는 사이트의 첫 인상에 의해 결정된다. 모두 재료의 성능에 따라 결정된다.

Q4: What are some architectural projects that inspired you regarding brick, tile, wood and/or glass? And why?

A: Agricultural green house. It acts as structure for controlling environment and just enhances growth in plants.

A: 농업용 온실. 환경을 통제하기 위한 구조 역할을 하며 식물의 성장을 향상시킨다.

Q5: Tell us about the materials you are interested in or want to use in your projects right now.

A: Nothing particular but I would like to use simple and easy materials as much as possible to make architecture.

A: 특별히 없지만 간단하고 쉬운 재료를 최대한 많이 쓰면서 건축을 만들고 싶다.

BRICK-Q1: Tell us about your favorite project that you used brick in or another architect's work - interior, facade, etc.

A: Melnikov House / Konstantin Melnikov

I was surprised to know a house, such as the Cylinder wall modern house with its many hexagonal shape windows was made of bricks, rather than reinforced concrete. At that time, I felt good because I could instantly comprehend the mystery of impressive design. Masonry could easily create curved wall plastically hexagonal shape windows with laid bricks by forty five degrees angles. The placement of the windows fits to a brick construction process and structural mechanics. I assume that Melnikov used masonry bricks because of financial constraints, but I would like to give high evaluation to his freewheeling design and effort which he put into small residence that he elevated the once sublimated, old masonry technique to a structure which embodies modernism in its composed structure and design.

A: 멜닉코브 하우스 / 콘슨탄틴 멜니코브
나는 육각형 모양의 창문이 많은 원통형 벽이 있는 현대식 주택이 철근 콘크리트가 아닌 벽돌로 만들어졌다는 사실에 놀랐다. 그 당시 나는 이 인상적인 디자인의 수수께끼를 바로 이해할 수 있었기 때문에 기분이 좋았다. 벽돌을 45도 각도로 배치함으로써 육각형 모양의 창문이 있는 유연한 곡선의 벽을 쉽게 만들 수 있었다. 창문의 배치는 벽돌 건설 과정과 구조 역학에 적합하다. 멜니코프가 재정적 제약 때문에 벽돌을 사용했을 거로 추측하지만, 나는 이 작은 주택에 넣은 자유분방한 디자인과 노력에 대해 높은 평가를 하고 싶다. 멜니코브는 한때 승화되고 오래된 벽돌 기술을 차분한 구조와 디자인을 통해 모더니즘을 구현하는 구조로 향상시켰다.

BRICK-Q2: What are the strengths and

Melnikov House ©Sergei Arssenev

weaknesses of brick?

A: Strength-It can become ruins.
Weakness-It can not be roofed.

A: 벽돌은 유적이 될 수 있지만 지붕으로는 쓸 수 없다.

TILE-Q1: Tell us about your favourite project that you used tile in or another architect's work - interior, facade, etc.

A: La maison de Jean Pierre Raynaud / Jean Pierre Raynaud
Raynaud completely covered the interior of his house with white tiles and created whole space as his artwork. New space stands as bulk material of accumulation when all materials were restored into a single tile scale. Thus, He represented how masonry joints play an important role in scale to fill newly-opened space. I was attracted by his artwork because it creates freedom in space that is confined by the restriction of surficial, divisible nature of tiles, transforming it into super-monomaniac space that compulsively creates a feeling of minimalism. Furthermore, the style of tiling itself attracts me more as his artwork.

A: 쟝 피에르 레노의 집 / 쟝 피에르 레노
레노는 흰색 타일로 집 내부를 완전히 덮고 공간 전체를 자기 작품으로 만들었다. 이 공간에 있던 모든 재료가 타일 한 장의 축척으로 복원될 때 새 공간은 축적의 대량자재가 된다. 따라서 그는 석조 조인트가 새롭게 열린 이 공간을 채우기 위해 축척 면에서 어떻게 중요한 역할을 하는지 보여준다. 나는 타일의 표면적이고 분열 가능한 특성으로 제한되는 공간에서 자유를 만들어 내면서 이를 강박적으로 미니멀리즘의 느낌을 주는 초 편집광적인 공간으로 변형시키는 그의 작품에 매료되었다. 게다가, 타일링 그 자체의 스타일은 그의 작품으로서 더욱 매력적이다.

TILE-Q2: What are the strengths and weaknesses of tile?

A: Strength-You feel like being in the water when you are in a tiled room. Weakness-It slips easily.

A: 타일로 된 방에 있으면 물 속에 있는 것 같다. 하지만 미끄러지기 쉽다.

192 Brick, Brick! What do you want to be?

GLASS-Q1: Tell us about your favourite project that you used glass in or another architect's work - interior, facade, etc.

A: Clean Room / Yuji Takeoka
Aside context in art, I like this artwork, because it simultaneously creates two spaces such as an empty space inside glass box and huddled imbrication space of spectators and other exhibitions, which created by reflection of glass exterior.

A: 클린 룸/ 타케오카 유지
예술적인 맥락을 제외하더라도 나는 이 작품을 좋아한다. 유리 상자 안의 빈 공간, 그리고 관중 및 다른 전시작품이 유리 외관에 반사되어 보이는 비늘 무늬 공간과 같은 두 개의 공간을 동시에 만들기 때문이다.

GLASS-Q2: What are the strengths and weaknesses of glass?

Sculptures by Jean Pierre Raynaud ©Jordiferrer

A: Strength-It is transparent material. Weakness-It is too strong and too suddenly united between inside and outside through glass without consciousness.

A: 유리는 투명한 재료이다. 하지만 너무 강하고 자각없이 내,외부가 갑자기 합쳐진다.

WOOD-Q1: Tell us about your favourite project that you used wood in or another architect's work - interior, facade, etc.

A: Gassho / Koji Kakiuchi
It is a small shelter created on the remnants of a concrete foundation of a home that was swept away by the 2011 Tsunami in northern Japan. It was entirely made of wood and designed in a single day of DIY construction. I wish to pay my greatest respects to this architect who bravely took action and creatively utilized characteristics of wood in this design.

A: 가쇼 / 카키우치 코지
가쇼는 2011년 일본 북부에서 쓰나미에 휩쓸려 간 집의 콘크리트 기초 잔해에 만들어진 작은 피난처이다. 이는 전적으로 나무로 만들어졌으며 DIY 건설 하루 만에 설계되었다. 용감하게 행동을 취하고 나무의 특성을 창조적으로 활용한 이 건축가에게 가장 큰 존경을 표하고 싶다.

WOOD-Q2: What are the strengths and weaknesses of wood?

A: Strength-It is easy for all aspect and has good touch feeling. Weakness-It burns easily.

A: 목재는 모든 면에서 다루기 쉽고 감촉이 좋지만 쉽게 타는 재질이다.

KimuraMatsumoto Architects office

Who is …?
Yoshinari Kimura
2003 Established Kimura Matsumoto Architects

Naoko Matsumoto
2003 Established Kimura Matsumoto Architects

Katsunobu Tasho
2003 Established katsunobu tasho / architecture

Masaki Kato
2013 Established masakikato

Q1: What is material to an architect
    (or to you)?

A: Material is an important element in any project that serves to translate the abstract image of architecture into a physical and tangible phenomenon. By focusing on harmonizing the originality and unique characteristics of both the design and materials, how to use those materials, or how to arrange that plan, can be changed. I think the physical and tangible nature of materials contributes to the creation of a freer and new architecture with higher versatility.

A: 어떤 프로젝트에서든지 재료는 건축의 추상적인 이미지를 물리적이고 유형적인 현상으로 변환하는 데 도움이 되는 중요한 요소이다. 디자인과 재료의 독창성과 고유한 특성을 조화시키는 데 초점을 맞춤으로써 재료를 사용하는 방법 또는 평면을 배열하는 방법을 변경할 수 있다. 나는 재료의 물리적, 유형적 특성이 매우 다양하고 자유로운 새 건축을 만드는데 기여한다고 생각한다.

Q2: Tell us about your favourite
    (or most often used) material and why.

A: **I am interested in materials which have a uniqueness to them.** For instance, bricks are manufactured goods which are produced under a carefully managed process, but the color of each baked brick differs slightly depending on the soil composition during the blending process. Those individual differences stand as a record of the conditions under which it was manufactured. I have a deep interest and faith in materials which show their

A: **나는 독특함이 있는 재료에 관심이 있다.** 예를 들어, 벽돌은 신중하게 관리되는 공정으로 생산되는 제품이지만, 구워진 벽돌의 색상은 혼합 공정 중 토양 조성에 따라 약간씩 다르다. 이러한 개별적인 차이는 제조 환경의 기록이 된다. 나는 외관에 역사가 나타나는 재료에 깊은 관심과 믿음이 있다.

Material is an important element in any project.
어떤 프로젝트에서든지 재료는 중요한 요소이다.

history in their appearance.

Q3: When do you decide the material during the design process and what is your criteria? (e.g. budget, client's preference, design concept, climate, etc.)

A: When I decide upon the materials depends on the project / varies from project to project. Sometimes I decide in the initial stages, but most of the time I do so during the design process. Occasionally I discover a new material by chance. In that case, I carefully examine its unique characteristics and try to enhance its usability. I always approach a new material with an open mind.

A: 내가 재료를 결정하는 때는 프로젝트에 따라 다르다. 가끔은 초기 단계에서 결정하지만, 대부분의 경우 설계 과정 중 결정한다. 가끔 우연히 새로운 재료를 발견하기도 한다. 이 경우 그 재료의 독특한 특징을 주의 깊게 살펴보고 유용성을 향상하려 노력한다. 나는 항상 열린 마음으로 새로운 재료에 접근한다.

Q4: What are some architectural projects that inspired you regarding brick, tile, wood and/or glass? And why?

A: I have been fascinated with many extraordinary examples of architecture, but on the other hand, learned many things from the ordinary scenery and anonymous buildings of daily life. For example, with buildings constructed on-site, I see the traces of skilled craftsman laying brick and

A: 나는 많은 특출난 건축물에 매료되어 왔지만, 다른 한편으로는 평범한 풍경과 일상생활에서 찾아볼 수 있는 익명의 건물에서 많은 것을 배웠다. 예를 들어, 현장에서 짓는 건물에서 숙련된 장인이 벽돌과 타일을 깔고 있는 흔

tile, and am always impressed with their careful and ingenious work in spite of the time-consuming process. **Of course, brick and tile by themselves are components of architecture, but if we look closer, the subtle differences created by the craftsman's placement of these features can result in a condition unique to that building.**

As to the knowledge we can acquire from the material, there is much to be learnt from the discourse between the concept and reality. Referring to the drawings from the Friedrichstrasse Office Building Project by Mies van der Rohe, as is often said, we see an opaque glass facade, even though we perceive glass as transparent. Sometimes, depending on the condition of the light, glass can look like a lead colored wall. Moreover, water stains are part of the properties of glass and yet rarely exist in the initial concept.

The white walls of the Todai-ji Daibutsuden in Nara, a familiar example of large timber architecture, appears from afar to be plaster but on closer inspection are infact painted wooden boards. I want to positively accept the gap between concept and reality - or the difference resulting

적이 보이고, 시간이 오래 걸리는 과정임에도 불구하고 신중하고 독창적인 작업에 항상 감명받는다. **물론 벽돌과 타일 자체가 건축의 구성 요소이지만, 자세히 살펴보면 장인이 재료를 배치할 때 미묘한 차이가 발생하여 그 건물에 고유한 상태가 생긴다.**

우리가 재료에서 얻을 수 있는 지식을 논하자면, 개념과 현실 사이의 담론에서 배울 것이 많다. 미스 반 데어 로에가 프리드리히스트라세 사무실 건물 프로젝트에서 그린 그림을 보자. 자주 언급되는 점으로, 우리는 보통 유리가 투명하다고 인식하지만 여기서는 불투명한 유리 외관이 보인다. 빛의 상태에 따라 유리는 때로 납색으로 보이기도 한다. 또한 물 얼룩은 유리의 특성 중 하나이지만 콘셉트 초기에 거의 고려되지 않는다.

일본 나라에 있는 도다이지 다이부쓰덴의 흰색 벽은 익숙한 큰 목재 건축물이다. 멀리서는 석고로 보이지만 더 자세히 살펴보면 실제로는 칠해진 나무 판자이다. 나는 이러한 개념과 현실 사이의 차이, 또는 물질의 특성으로 인한 차이를 긍정

Todai-ji Daibutsu-den in Nara ©Nekosuki

from the property of a material. Perhaps we can be emboldened and enriched by what they have to offer. This is because I think that this attitude could help to reaffirm our common sense; broadening the potential uses of materials and enhance the resulting architecture.

적으로 받아들이고 싶다. 어쩌면 우리는 이런 차이로 대담해지고 농축될 수 있을지도 모른다. 이러한 태도가 우리의 상식을 재확인하고, 재료의 잠재적인 사용을 확대하고, 결과되는 건축을 향상하는 데 도움이 될 수 있다고 생각한다.

Q5: Tell us about the materials you are interested in or want to use in your projects right now.

A: I am interested in all materials, and the various different ways to use them. I have no preference between natural

A: 나는 모든 재료와 이를 사용하는 다양한 방법에 관심이 있다. 천연 재료든 인공 재료

and artificial materials. The subtle hue of a brick, gradation in the glaze of a tile, scattered fibers within FRP Board or the grain of timber all bear the same meaning - as it is a material's history and unique character that interests me. Perhaps you could say I value uniqueness? Natural conditions, production methods, traces of the craftsman's intentions and skill - qualities that we as architects are unable to control; materials which bear these features are ones I have a deep interest in.

든 특별히 선호하지 않는다. 나에게 벽돌의 미묘한 색조, 타일 유약의 그라데이션, FRP 보드의 흩어진 섬유질, 혹은 목재의 입자 모두 똑같다. 나는 재료의 역사와 독특한 특성이 흥미롭기 때문이다. 어떻게 보면 나는 독특함을 중시한다고 할 수도 있겠다. 자연조건, 생산 방법, 장인의 의도와 기술의 흔적 등 건축가로서 우리가 통제할 수 없는 특성 말이다. 이러한 특징을 지닌 재료가 내가 깊이 관심이 가는 재료이다.

BRICK-Q1: Tell us about your favorite project that you used brick in or another architect's work - interior, facade, etc.

A: Brick masonry architectural projects that I admire are too numerous to mention. Needless to say, the interesting feature of brickwork is that it is a fundamental unit which when combined can create a wall, floor, ceiling and evening furniture such as chairs and tables. I think the best example of this is "Church of St Peter" by Sigurd Lewerentz.

A: 내가 동경하는 벽돌 건축 프로젝트는 너무 많아서 하나만 언급하기 너무 어렵다. 벽돌의 가장 흥미로운 특징은 벽, 바닥, 천장 그리고 의자나 탁자 같은 모든 가구를 만들 수 있는 기본 단위라는 것이다. 이것의 가장 좋은 예는 Sigurd Lewerentz의 "성 베드로 교회"라고 생각한다.

BRICK-Q2: What are the strengths and

weaknesses of brick?

A: During the modernization in Japan, brick masonry buildings were constructed to prevent fires in urban areas crowded with timber houses. In the project "Another base in G" we tried to reduce the sound transmission by focusing on the inherent sound insulating qualities resulting from the weight of brick.

Both of these cases were positively affected by the characteristics of brick masonry.
As I mentioned above, although a brick's hardness can make it difficult to process, its form and size (designed to be carried in the hand) make it possible to form various shapes. This ambiguity is something I find attractive and the balance between a material's strengths and weaknesses is something I pay close attention to.

For example, by using brick masonry as a structural wall we lose the ability to create openings at will. However rather than see this as a negative, we can except this and celebrate its unexpected results.

Not just in the case of brickwork, but

A: 일본의 근대화 과정에서 목조 가옥의 화재를 예방하기 위해 벽돌 건물을 지었다.
"Another base in G" 프로젝트에서 우리는 벽돌의 무게로 인한 고유한 방음 특성에 초점을 맞추어 음향 전달을 줄이려고 했다.

이 두 사건 모두 벽돌 조적의 특성에 의해 긍정적으로 영향을 받았다.
내가 위에서 언급했듯이, 벽돌의 경도는 처리를 어렵게 할 수 있지만, 벽돌의 형태와 크기는 다양한 모양을 만드는 것을 가능하게 한다. 이 모호함은 내가 매력적이라고 생각하는 것이고 재료의 강점과 약점 사이의 균형은 내가 세심하게 주의를 기울이는 것이다.

예를 들어, 벽돌을 구조 벽으로 사용함으로써, 우리는 마음대로 개구부를 만들 수 있는 능력을 잃게 된다. 하지만 이것을 부정적으로 보기보다는, 그것을 제외시키고 다른 예상치 못한 결과를 축하할 수 있다.

벽돌의 경우뿐만 아니라, 어떤 재료에 대해서도, 그것의 장점과 단점을 이해하고 그것들을 가지고 작업하려고 시도하는

with any material, I think it is important to understand and accept both it's strengths and weaknesses and attempt to work with them.

것이 중요하다고 생각한다.

Another base in 'G'

LANDÍNEZ+REY architects

Who is ...?

LANDINEZ + REY is an architectural practice co-founded in 2000 and based in Madrid(Spain) by the architects David Landinez González-Valcácel(Madrid, 1973) and Mónica González Rey(Paris, 1973). Both are formed as M.Arch (1999) in ETSAM-UPM(Faculty of Architecture of the Polytechnical University of Madrid, UPM) and both are also graduated as Building Engineers by UEM-Madrid(2013). David also is M.Arch in Efficient Buildings and Rehabilitation by UEM (Universidad Europa de Madrid) and Mónica has also postgraduated studies in Analysis and Real Estate Management by Colmillas University (ICAI-ICADE)

Q1: What is material to an architect
(or to you)?

A: **The architectural space arises only in the dialogue between the material and the form it generates:** that's why the material is much more than the negative of the architectural space.
The material talks about the character and the intention of the architectural project in its correct (or incorrect) constructive disposition. The construction is the systematization of the disposition of the materials, and the architecture systematizes its use from its own disciplinary parameters.

A: 건축 공간은 재료와 재료가 생성하는 형태 사이의 대담을 통해서만 생긴다. 재료가 그저 건축의 네거티브 스페이스 이상인 이유이다. 재료는 올바른 (또는 그른) 건설적인 사용법에 대한 프로젝트의 성향과 의도에 관해 이야기한다. 건설은 재료의 성질을 체계화하고, 건축은 분야의 한도 내에서 재료의 사용을 체계화한다.

Q2: Tell us about your favourite
(or most often used) material and why.

A: We have no predilection for any material, but we reject the use of any that lose their dignity for an incorrect constructive use: imitations and plating we would only use them as a possible joke. The construction by itself is not architecture, but as architects (an also building engineer) we respect this discipline as we respect the concept of **FIRMITAS** taught by Marcus Vitruvius

A: 우리는 어떤 재료에도 편향이 없지만, 잘못된 건설 방식으로 재료의 존엄성을 떨어뜨리는 것은 거부한다. 모방재와 도금은 농담으로만 쓸 것이다. 건설 자체로만은 건축이라 할 수 없지만, 건축가(그리고 건축 엔지니어)로서 우리는 마르쿠스 비트루비우스(Marcus Vitruvius)가 말했던 **FIRMITAS(견고함)**의 개념을 존중하기 때문에 이 분야와 사

and, in this sense, we respect this discipline and therefore the adequacy of the material to use (**UTILITAS**).

Q3: What are some architectural projects that inspired you regarding brick, tile, wood and/or glass? And why?

A: The arrival to architecture is not only made from the analysis of its precedents. Of course, we trust in the arrival of architecture from the knowledge of history and those concepts that define it as a discipline, but also from the economy, technology or other transversal disciplines.

Q4: Tell us about the materials you are interested in or want to use in your projects right now.

A: The materials that interest us are those that are capable of accompanying our proposals and ideas. We put geometry in opposition with topology. Topology is understood as the way the parts of something are organized or connected.
If Topology speaks us about continuity, but also about other properties

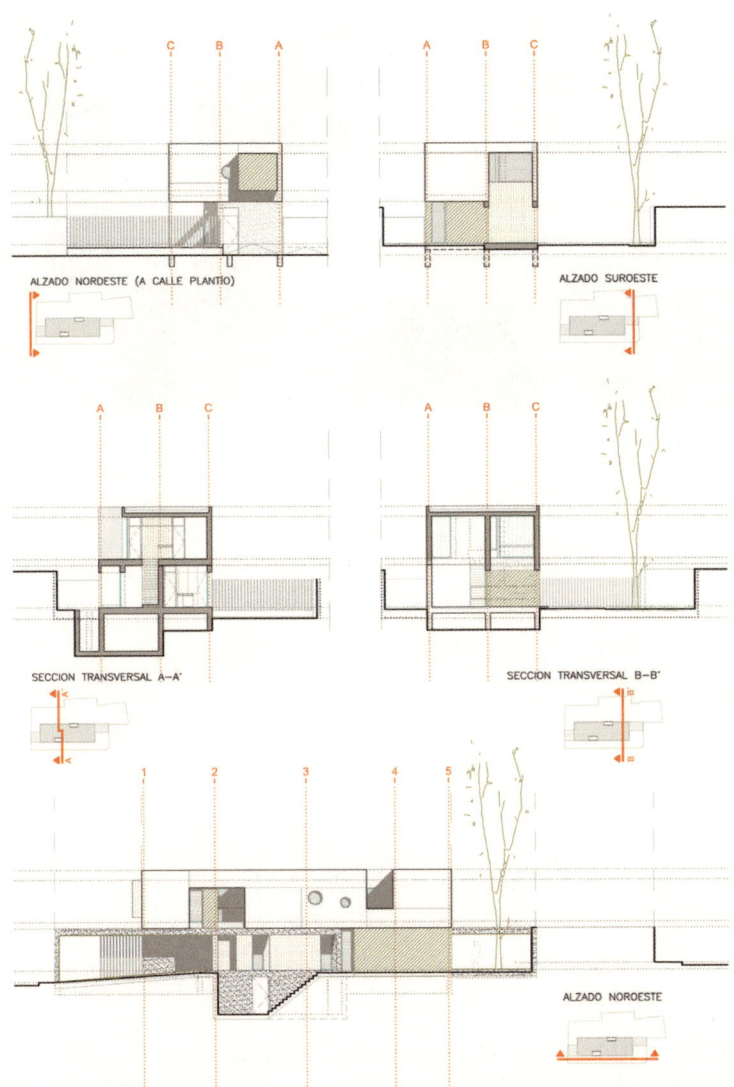

the architecture systematizes its use from its own disciplinary parameters.

건축은 분야의 한도 내에서 재료의 사용을 체계화한다.

DEMITAS & UTILITAS casa iA

regardless of the size or shape of figures, construction forces us to geometry: dimension drift us to the scale.
We set geometry in opposition with topology, and pit construction and scale against continuity and time.

라, 사물의 크기나 모양과 관계없이 다른 다양한 속성도 논할 수 있다. 건설은 기하학을 강요하고, 크기는 축척으로 표류하게 만든다.
우리는 토폴로지와 기하학을 대비시키고 건설과 축척을 연속성과 시간에 맞붙인다.

BRICK-Q1: Tell us about your favorite project that you used brick in or another architect's work - interior, facade, etc.

A: In the office we appreciate any project that investigates the possibilities of each material to get its architectonical premises. In fact, i do not remember a project in which we have not used bricks and we never make a difference in their disposition for being inside or outside. That's why we, as an architectural office, have grown up accompanied by

A: 우리 사무소에서는 각 재료가 건축학적 전제를 얻을 가능성을 조사하는 모든 프로젝트를 환영한다. 사실, 내 기억에 우리가 벽돌을 사용한 프로젝트 중 내부든 외부든 그 사용 방식에 변화를 주지 않았던 프로젝트는 없었다. 우리가 건축 사무소로서 벽돌의 사용, 기술 및 표현력에 대한 진보적인 지

our progressive knowledge in its use, technology and expressiveness.

식과 함께 성장해온 이유이다.

BRICK-Q2: What are the strengths and weaknesses of brick?

A: It is evident that the use of any construction system is intimately linked to what is its price. In the case of brick this depends directly on the price of labor and the tradition of using the material in each country. In Spain it is a technology with a huge industrial experience and technological development, which has led to a constant experimentation by architects.
The strength of the brick is its expressive capacity and experimentation capabilities, as well as its local tradition. It weakness is only the result of its misuse.

A: 어떤 건설 방식을 사용하든 가격이 얼마인지와 밀접하게 관련되어 있다는 점은 명확하다. 벽돌의 경우 이것은 각 국의 인건비와 전통 재료에 직접 연관이 있다. 스페인에서 벽돌은 거대한 산업 경험과 기술의 발달을 갖춘 기술이므로, 건축가가 지속해서 여러가지로 실험해 올 수 있었다. 벽돌의 장점은 표현 능력과 실험성뿐만 아니라 이 지역의 전통인 점이다. 단점이 있다면 단지 오용의 결과일 뿐이다.

TILE-Q1: Tell us about your favourite project that you used tile in or another architect's work - interior, facade, etc.

A: We still fascinated by the use of the tile in the fifth facade that results from the project of Mercat de Santa Caterina

A: 우리는 2005년 엔릭 미라예스(Enric Mirallas)와 베네데타 타글리아부에(Benedetta Tagliabue)가 스페인 바르셀로

(Barcelona, Spain) by Enric Miralles and Benedetta Tagliabue, 2005.

나에 있는 산타 카테리나 시장 (Mercat de Santa Caterina)의 다섯 번째 파사드에 타일을 사용한 방식에 여전히 매료되어 있다.

Market Santa Caterina ©Tony Hisgett

TILE-Q2: What are the strengths and weaknesses of tile?

A: The tile industry in Spain is one of the most powerful in the world, with cutting-edge technological development. Undoubtedly, the good use in its tradition, together with this representation of the future, constitutes its best potential.

A: 스페인의 타일 산업은 최첨단 기술 개발로 세계에서 가장 뛰어나다고 말할 수 있다. 말할 것도 없이, 최첨단 기술이라는 미래의 표현과 전통적인 활용법이 만나 최고의 잠재력을 만든다.

GLASS-Q1: Tell us about your favourite project that you used glass in or another architect's work - interior, facade, etc.

A: Undoubtedly our references for glass architecture are those of the history of architecture. In the first place, the radical and wise use of glass in the Maison de Verre by Pierre Chareau or the Pavilion by Bruno Taut for the Deutcher Werkbund in 1914: these are architectural lessons for glass use. In advance, the architecture of Mies or, in 1949, the Glass House of Philip Johnson. that seem to evolve towards the use of glass, today, by Kazuyo Sejima.

A: 유리 건축에 대한 우리의 선례는 의심할 여지 없이 건축 역사 안에서 찾을 수 있다. 우선, 1914년 피에르 샤로(Pierre Chareau)의 메종 드 베르(Maison de Verre)와 브루노 타우트(Bruno Taut)의 독일 베르크분트(Deutcher Werkbund)를 위한 파빌리온의 급진적이고 현명한 사용법은 유리를 쓰는데 필요한 건축적 교훈이다. 이 시대 이후로, 미스 반 데어 로에(Mies van der Rohe)의 건축이나 1949년에 지어진 필립 존슨(Philip Johnson)의 글래스 하우스(Glass House)에서 오늘날 세지마 카즈요의 유리 사용법까지 발전한 것 같다.

Taut Glass Pavilion

Our Train Station in Rivas-Futura (Madrid, Spain) is definitely a glass box arranged on a podium and protected with a shade covering floating above

스페인 마드리드에 있는 우리 사무소의 리바스-푸투라 (Rivas-Futura) 기차역은 확실히 연단에 배열된 유리 상자

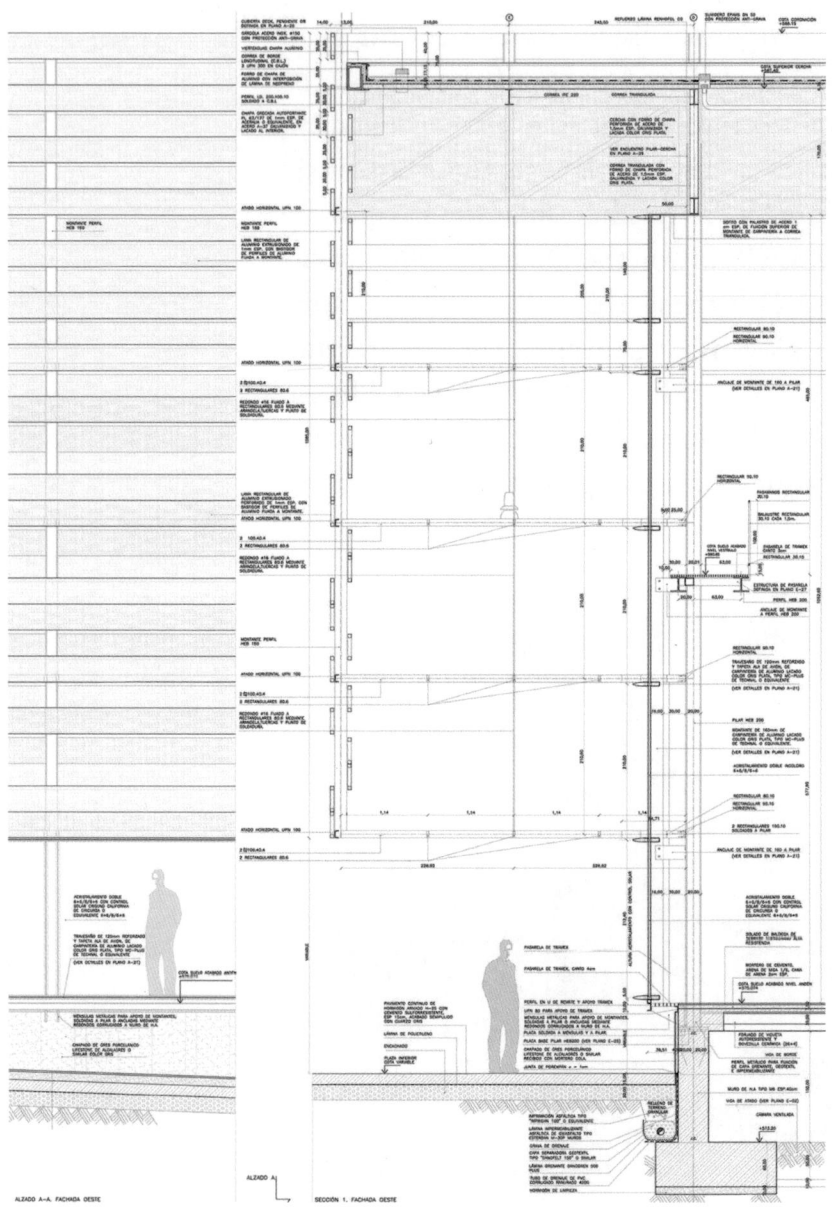

Train Station in Rivas-Futura
©MIGUEL DE GUZMÁN

it, according to our climate. By linear aluminum louvers in front of the glazing, a bioclimatic shade covering is created.
It offers protection against west solar direct radiation avoiding overheating and provides reflected natural lighting inside the spaces avoiding glare. Also, a covered exterior area where the external stairs between public spaces is provided.

로, 여기 기후에 맞춰 유리 위에 떠 있는 가리개로 보호되어 있다. 유리 앞의 선형 알루미늄 루버가 친환경 그늘 덮개를 만든다. 이 루버는 또 서쪽 태양의 직사광을 막아 과열을 방지하고, 눈부심을 막으면서 내부에 빛을 반사하여 자연조명 효과도 만든다. 또한 공공 공간 사이 외부 계단이 있는 곳에 지붕이 있는 외부 영역도 만든다.

GLASS-Q2: What are the strengths and weaknesses of glass?

A: The strengths and weaknesses of the use of glass in architecture have been linked to their intrinsic values around transparency, or not, or to their light-responsive properties.
Today the classic discourse of the use of glass in architecture has been added in an essential way its response in relation to the energetic factors.

A: 건축에서 유리 사용의 장단점은 투명성을 둘러싼 본질적인 가치 또는 빛에 반응하는 특성과 관련이 있을지도, 없을지도 모르겠다.
오늘날 건축에서 유리 사용에 대한 고전적인 담론은 에너지 요인이 필수적으로 추가되었다.

WOOD-Q1: Tell us about your favourite project that you used wood in or another architect's work - interior, facade, etc.

A: One of the lesser-known episodes of Le Corbusier's work refers to the set of buildings that the architect would make for his personal vacations in Roquebrune-Cap Martín, his place of retirement in the French Riviera.

A: 르 코르뷔지에(Le Corbusier)의 덜 알려진 작품 중 하나로, 프랑스 리비에라(Riviera)에 있는 로크브륀느-카프마르탱(Roquebrune-Cap Martín)에 자신의 사적인 휴가를 위해 만든 건물이 있다. 그는 나중에 이곳으로 은퇴했다.

WOOD-Q2: What are the strengths and weaknesses of wood?

A: We must use the materials of our

A: 우린 우리 주변 환경의 재

environment: those that the earth offers us in its proximity. We do not find any weakness in the use of wood in architecture. Any weakness in its use is purely cultural or linked to the absence of this material.

료를 사용해야 한다. 지구가 우리에게 주는 우리 주변에 있는 재료 말이다. 우리는 건축에서 나무를 사용하는데 어떤 단점도 못 찾았다. 목재를 사용하는데 단점이 있다면 순전히 문화적인 이유거나 재료의 부재와 관련이 있다.

"WE REJECT THE USE OF ANY THAT LOSE THEIR DIGNITY FOR AN INCORRECT CONSTRUCTIVE USE."

Cabanon ©Tangopaso

M artı D Mimarlık

Who is ...?

M artı D Mimarlık was founded in 1987 in Izmir by Metin Kılıç and Dürrin Süer. They design various types of projects in various scales such as residential, commercial, healtcare, educational and urban design. With their intention that unites academic and practical skills, they contribute to today's architecture culture.

Q1: What is material to an architect (or to you)?

A: Material is determinitve for forming building's /space's identity/soul and for their existence.

A: 재료는 건물 및 공간의 정체성과 영혼을 형성하고 그 존재를 위해 결정적이다.

Q2: Tell us about your favourite (or most often used) material and why.

A: We use exposed concrete frequently as structural system. It's an easy material to form and it's suitable for texture testing. Also it is economical and used widespreadly. We prefer stone as cladding material because, its durability to exterior effects, expands building's aging without losing its quality.

A: 우리는 구조 시스템으로 노출된 콘크리트를 자주 사용한다. 형성하기 쉽고 질감을 실험해보기에 적합하다. 또한 콘크리트는 경제적이며 널리 사용된다. 우리는 클래딩 재료로는 돌을 선호하는 데 외부 환경에서 내구성이 좋고, 품질을 잃지 않으면서 건물의 수명을 늘리기 때문이다.

Q3: When do you decide the material during the design process and what is your criteria? (e.g. budget, client's preference, design concept, climate, etc.)

A: At design process, building's location, function, budget, manifacturing process affect material choice. **Our general intention is to use material as it natural state.** For instance, we don't prefer to

A: 설계 과정에서 건물의 위치, 기능, 예산, 제조 공정이 재료 선택에 영향을 미친다. **대체로 재료를 자연적인 상태로 사용하려고 한다.** 예를 들어, 우리는 돌이나

use stone or wood looking materials. **We prefer materials that minimise maintenance costs.**

나무처럼 보이는 재료를 선호하지 않는다. **우리는 보수 비용을 최소화하는 재료를 선호한다.**

Q4: What are some architectural projects that inspired you regarding brick, tile, wood and/or glass? And why?

A: Brick is a sustainable material, due to being soil-based material. Being modular materials eases construction. Especially, impressive examples can be seen where economical limitations is a big issue. SCC_LAS TOSCAS by Baarqs, NEW ACCESS TO GIRONELLA'S HISTORIC CENTRE by Carles Enrich can be considered as good examples for brick use.

A: 벽돌은 토양을 기반으로 한 재료이기 때문에 지속 가능한 물질이다. 모듈형 재료가 있으면 건설이 쉬워진다. 특히 경제적 한계가 큰 문제인 경우 인상적인 사례를 볼 수 있다. 바아크(Baarqs)의 SCC 라스 토스카스 (SCC_LAS TOSCAS)와 칼스 엔리히 지메네즈(Carles Enrich Giménez)의 지로넬라 역사 센터 (Gironella Historic Centre)에서 볼 수 있는 새로운 접근은 벽돌 사용의 좋은 예로 간주할 수 있다.

Q5: Tell us about the materials you are interested in or want to use in your projects right now.

A: Recycled materials are attracting my attention but we haven't used them in a project yet.

A: 전에도 말했듯이, 우리가 특히 선호하는 재료는 없다. 건축 자재는 사용하는 데에 가치를 부여하는 것이 중요하다.

BRICK-Q1: Tell us about your favorite

project that you used brick in or another architect's work - interior, facade, etc.

A: Brick is used as cladding material in Asmabahçeler Residences. For many users, perception of brick is considered as sympatic. Brick is consistent with notion of "home". For this reason we have used brick in this project. Furthermore, with developing application techniques and technology, it provides heat insulation.

A: 벽돌은 아스마바흘라 주택 (Asmabahçeler Residences)에서 클래딩 재료로 사용했다. 많은 이가 벽돌을 호의적으로 인식하고, 벽돌은 '집'이라는 개념과 일치한다. 이런 이유로 우리는 이 프로젝트에서 벽돌을 사용했다. 또한 벽돌은 건설 기법과 기술의 발달로 이제 단열재 역할도 한다.

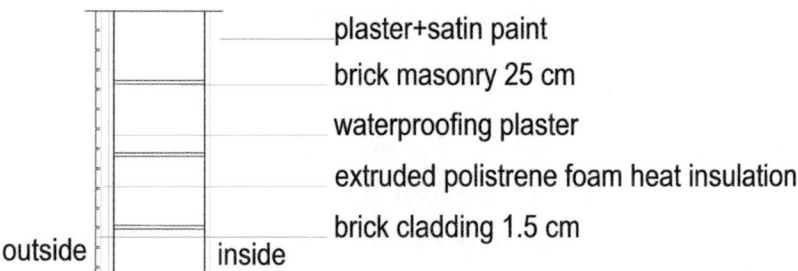

Asmabahçeler Residences-System Detail

BRICK-Q2: What are the strengths and weaknesses of brick?

A: Widespread use of brick enables builders to find it, also it is economical and builders are familiar with it.

A: 벽돌의 광범위한 사용으로 건설업자가 재료를 찾기 쉬우며 경제적이고 현장 업자에게 익숙하다.

TILE-Q1: Tell us about your favourite project that you used tile in or another architect's work - interior, facade, etc.

A: We use tiles generally at wet spaces like bathrooms, wc due to its hygenic and water-proofing characteristics. In Automative Exporters Vocational High school, we have used a large dimensioned tile (100x300cm) for interior wall cladding.

A: 우리는 일반적으로 위생 및 방수 특성 때문에 화장실과 같은 젖은 공간에서 타일을 사용한다. 우리는 자동차 수출자 직업 고등학교의 내부 벽 클래딩에 치수가 큰 타일 (100x300cm)을 사용했다.

TILE-Q2: What are the strengths and weaknesses of tile?

A: Characteristic of being modular material is useful for design process but, seams and joints may causes some hygiene problems. However, larger sizes of tiles have been produced for a few years and it minimises hygiene problems while increasing need of precise work at construction process.

A: 모듈형 재료의 특성은 설계 과정에는 유용하지만 이음매와 조인트가 위생상 문제를 일으킬 수 있다. 하지만 지난 몇 년간 더 큰 크기의 타일은 생산되면서 위생 문제는 최소화했지만, 건설 과정에서 정밀한 작업의 필요성이 높아졌다.

GLASS-Q1: Tell us about your favourite project that you used glass in or another architect's work - interior, facade, etc.

A: Izmir Chamber of Geological Engineers is a small scale project that we have designed on a narrow parcel. It is located on an attached parcel. Services, circulation elements are located at the back side of the parcel which enables building to get light in from one facade of it. For maximising use of natural light in the building, glass curtain wall was used on that facade. In order to, reflect institutional identity, topographical references are used for geometry of the facade.

A: 지질공학자 협회의 이즈미르 회의소는 우리가 좁은 구획에 설계한 소규모 프로젝트이며 부착된 구획에 있다. 서비스, 순환 요소는 구획의 뒷쪽에 있어 건물은 파사드 하나에서 채광을 받는다. 이 채광을 최대한 활용하기 위해 그 파사드에 유리 커튼월을 사용했다. 이 협회의 정체성을 반영하기 위해 파사드의 기하학에 지형학적 요소를 사용했다.

GLASS-Q2: What are the strengths and weaknesses of glass?

A: Glass is advantageous due to its transparent characteristics. However, bad heat control and sound insulation characteristics are improved by technological developments. Layered use of glass with air gaps and coating it by various materials, improves its heat insulation capacity. As a result use of glass becomes more easier and widespread.

A: 유리는 투명한 특성이 장점이지만 부족했던 단열 및 방음 특성은 기술적 개발로 향상했다. 유리를 간격을 두고 층층으로 사용하고 다양한 재료로 코팅하면 단열성이 향상된다. 결과적으로 유리를 사용하기 더 쉬워지며 더 널리 보급된다.

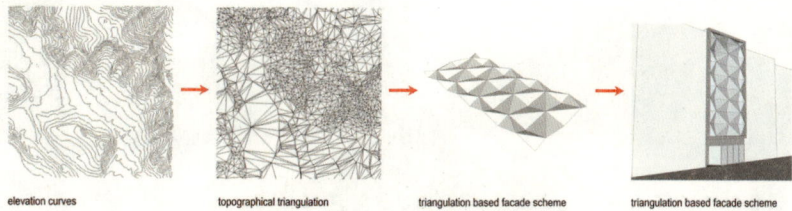

elevation curves   topographical triangulation   triangulation based facade scheme   triangulation based facade scheme

modostudio

Who is ...?
　modostudio | cibinel laurenti martocchia architetti associati, located in Rome, is a multidisciplinary practice of architecture, urban planning and industrial design.

Q1: What is material to an architect (or to you)?

A: **A material is the essence of an architectural project.** The quality of an architecture has a lot to do with the materials used for building it and it cannot be separated from the material. The material used in a building can represent the tradition of a site, the culture of a people.

A: **재료는 건축 프로젝트의 본질이다.** 건축의 품질은 건물에 사용되는 재료에 크게 좌우되며 재료와 분리할 수 없다. 건물에 사용된 재료는 사이트의 전통과 그곳 사람들의 문화를 나타낼 수 있다.

Q2: Tell us about your favourite (or most often used) material and why.

A: I don't have a favourite material. In our project we use all kind of materials, Stone, Steel, plastic, wood. We use the material as a way to interprete a building and a site. We don't have any preferences, but we think is important to use a materials preserving and showing its quality.

A: 특히 좋아하는 재료는 없다. 우리는 프로젝트에서 돌, 강철, 플라스틱, 나무 같은 모든 종류의 재료를 사용한다. 우리는 재료를 건물과 사이트를 해석하는 방법으로 사용한다. 선호하는 재료는 없지만, 재료의 특성을 보존하고 보여주는 것이 중요하다고 생각한다.

Q3: When do you decide the material during the design process and what is your criteria? (e.g. budget, client's preference, design concept, climate, etc.)

A: In our approach we use the material

A: 우리의 접근 방식은 일반적

The material used in a building can represent the tradition of a site, the culture of a people.

건물에 사용된 재료는 사이트의 전통과 그곳 사람들의 문화를 나타낼 수 있다.

Intecs headquarters ©JULIEN LANOO

Intecs headquarters ©JULIEN LANOO

Intecs headquarters ©JULIEN LANOO

226 Brick, Brick! What do you want to be?

generally as an important element of the architecture. We can use the material as a link with the culture of a particular site where the building is located, or we can use the material as a inspiring element of the architecture. We think about materials since the very beginning of our design process: materials, space, relation with the site are our most important inspiration points.

으로 재료를 건축의 중요한 요소로 사용한다. 재료를 건물이 있는 특정 장소의 문화와의 연결 고리로 사용하거나 건축적 영감을 주는 요소로 사용할 수 있다. 우리는 디자인 과정의 시작부터 재료에 대해 생각하며 재료, 공간, 사이트와의 관계는 가장 중요한 영감 포인트이다.

Q4: Tell us about the materials you are interested in or want to use in your projects right now.

A: As I told before we don't have a preferred material. **Is important to give to construction material a value in using it.**

A: 전에도 말했듯이, 우리가 특히 선호하는 재료는 없다. **건축 자재는 사용하는 데에 가치를 부여하는 것이 중요하다.**

BRICK-Q1: Tell us about your favorite project that you used brick in or another architect's work - interior, facade, etc.

A: Rome, the city where I've been born, where I live and work is truly inspiring for the use of the brick, starting from the ancient roman ruins. For sure, my preferred brick building is the Oratorio

A: 내가 태어났고, 살고 일하고 있는 도시인 로마는 고대 로마 유적지에서부터 벽돌 사용에 진정으로 영감이 넘친다. 내가 단연코 선호하는 벽돌 건물은 위대한 프란체스코

dei Filippini, from the great Francesco Borromini. An example of great capacity in construction techniques linked with an incredible design freedom.

보로미니의 오라토리오 데이 필리피니이다. 이 건물은 놀라운 디자인적 자유와 연계된 건설 기술의 훌륭한 가능성을 보여준다.

BRICK-Q2: What are the strengths and weaknesses of brick?

A: The brick is a material which is showing a strong consistence, but through its particular dimension and quality in able to create incredible variety of architectural structures, from the most simple ones to complex geometries.

A: 벽돌은 강한 일관성을 지닌 재료이지만, 그 특정한 크기과 특성을 통해 가장 단순한 것에서부터 복잡한 기하학에 이르기까지 놀랍도록 다양한 건축 구조를 만들 수 있다.

Seaside Single House

TILE-Q1: Tell us about your favourite project that you used tile in or another architect's work - interior, facade, etc.

A: Tiles make me remind the beautiful Amalfi coast where there is a great tradition of Ceramics artworks and tiles. In the small village of Vietri there is a beautiful building designed by Paolo Soleri the architect founder of Arcosanti (Arizona). The building called "Ceramica Solimene"is a beautiful free shaped volume cladded with traditional coloured ceramics tiles.

A: 타일은 세라믹 예술작품과 훌륭한 타일 전통이 있는 아름다운 아말피 해안을 상기시킨다. 비에트리라는 작은 마을에는 건축가이자 아르코산티(애리조나)의 설립자인 파올로 솔레리(Paolo Soleri)가 디자인한 아름다운 건물이 있다. 세라미카 솔리메네라고 불리는 이 건물은 전통적인 색의 도자기 타일로 덮인 자유로운 형태의 아름다운 볼륨이다.

TILE-Q2: What are the strengths and weaknesses of tile?

A: Tiles make possible to have incredible coloured surfaces and give a lot of freedom to assembly them.

A: 타일로 놀라운 색상이 가능하며 이를 아주 자유롭게 조합할 수 있다.

GLASS-Q1: Tell us about your favourite project that you used glass in or another architect's work - interior, facade, etc.

A: An architectural icon that was

A: 지어지지 않은 건축 아

never build is the Mies van der Rohe Glass Skyscraper (1922). Mies van der Rohe was one of the first architect to understand the poetics of glass and to bring this material to extreme solutions.

이콘으로 미즈 반 데어 로에(Mies van der Rohe)의 유리 고층건물(Glass Skyscraper, 1922)이 있다. 미즈 반 데어 로에는 유리의 시학을 이해하고 이 재료를 극한으로 발전시킨 최초의 건축가 중 한 명이다.

GLASS-Q2: What are the strengths and weaknesses of glass?

A: The strengths of the glass is its transparency, a characteristic which can become a weakness too. Glass is a material that has to be used with extreme care. It can bring lightness but only if carefully technological detailed.

A: 유리의 장점은 투명성이지만, 이는 단점이 될 수도 있다. 유리는 극도로 주의를 기울여 사용해야 하는 재료이며 신중하게 기술적으로 디테일을 디자인해야만 우아함을 지닐 수 있다.

WOOD-Q1: Tell us about your favourite project that you used wood in or another architect's work - interior, facade, etc.

A: The 17th century wooden Farnese theater in Parma is an incredible masterpiece. It was built as a private theater for Farnese family and it is an example of detailed artcrafts.

A: 파르마에 있는 17세기 목조 파르네세 극장은 놀라운 걸작이다. 파르네세 일가를 위한 개인 극장으로 지어졌으며 상시한 공예의 좋은 예이다.

WOOD-Q2: What are the strengths and

weaknesses of wood?

A: The wood is a material that can be used in various situation. It can be a structural construction material, it can be used as a decorative material. Wood is very flexible and has a strong link to the nature.

A: 나무는 다양한 상황에서 사용할 수 있는 재료이다. 구조용 건설 자재가 될 수도 있고, 장식용 재료가 될 수도 있다. 나무는 매우 유연하며 자연과 강한 연관성이 있다.

"THE MATERIAL TALKS ABOUT THE CHARACTER AND THE INTENTION OF THE ARCHITECTURAL PROJECT IN ITS CORRECT (OR INCORRECT) CONSTRUCTIVE DISPOSITION."

Mork-Ulnes Architects

Who is ...?
About Casper Mork-Ulnes

Norwegian born, Casper Mork-Ulnes was raised in Italy, Scotland and the United States, which has brought a broad perspective to his eponymous firm's work. In 2015, Casper was named one of "California's finest emerging talent" by the American Institute of Architects California Council. He was selected by the Norwegian National Museum as one of "the most noteworthy young architects in Norway" with the exhibit "Under 40. Young Norwegian Architecture 2013." Casper holds a Master of Architecture from Columbia University and a Bachelor of Architecture from California College of the Arts.

Q1: What is material to an architect (or to you)?

A: If used well - a material can give life to a building - and at the same time set the design parameters of the project.

A: 잘 사용한다면 재료는 건물에 생명을 주는 동시에 프로젝트의 디자인 사항을 결정한다.

Q2: Tell us about your favourite (or most often used) material and why.

A: Wood is perhaps the most versatile building material. It can be shaped to almost any form, it can support a building or be the finish material of every surface. **Wood has very few limitations and evokes the human senses in more different ways than perhaps any other material.** It is a very humble basic material and can be very complex at the same time, and its patina improves over time.

A: 나무는 아마도 가장 다용도의 건축 자재일 것이다. 거의 모든 형태로 형성될 수 있으며, 건물을 지탱할 수도, 모든 표면의 마감재가 될 수도 있다. **나무에는 한계가 거의 없으며 아마 다른 어떤 재료보다 더 다양한 방식으로 인간의 감각을 불러일으킨다.** 매우 소박한 기본 재료인 동시에 매우 복합적일 수 있고 시간이 지남에 따라 고색이 향상된다.

Q3: When do you decide the material during the design process and what is your criteria? (e.g. budget, client's preference, design concept, climate, etc.)

A: Material choice is affected by many things – but sometimes it can be a very natural choice. In the case of the Mylla

A: 재료 선택은 많은 것에 영향을 받지만 가끔 아주 자연스럽게 결정될 때도 있다. 밀라 프로젝트의 경우 우리는 목재

Mork-Ulnes Architects 233

project we didn't consider any other materials than wood – the only question was the species and finish. That said in other cases, the material choices can be a complex process where context, budget, climate, maintenance etc. all need to be weighed before making the right selection.

Q4: What are some architectural projects that inspired you regarding brick, tile, wood and/or glass? And why?

A: Difficult question but perhaps Maison de Verre by Pierre Chareau for pushing glass as a material in a new way. And vernacular timber structures - like the Norwegian stave churches still surviving after being built almost 1000 years ago - for their resilience and continued beauty as a material with a developing patina.

Q5: Tell us about the materials you are interested in or want to use in your projects right now.

A: We are exploring concrete in two new projects. Both on its own, as well as in conjunction with glass and wood.

이외의 다른 재료는 고려하지 않았다. 유일한 문제는 나무의 종과 마무리였다. 그렇지만 대부분의 경우 올바른 재료 선택을 하기 전에 콘텍스트, 예산, 기후, 유지 보수 등 모든 것을 재봐야 하는 복잡한 과정일 수 있다.

A: 어려운 질문이지만 피에르 샤로(Pierre Chareau)의 메종 드 베르(Maison de Verre)일지도 모르겠다. 유리를 건축 자재로서는 새로운 방식으로 썼기 때문이다. 그리고 시간이 지날수록 더해가는 고색을 지닌 재료로, 복원력과 지속되는 아름다움을 지닌 노르웨이의 통널 교회 같은 지역 목재 구조를 고르겠다. 이는 지어진 지 거의 천년이 지났음에도 아직도 서 있다.

A: 우리는 새로운 두 개의 프로젝트에서 자체적으로 콘크리트를 탐구해보고 있다. 그

Wood is perhaps the most versatile building material.
나무는 아마도 가장 다용도의 건축 자재일 것이다.

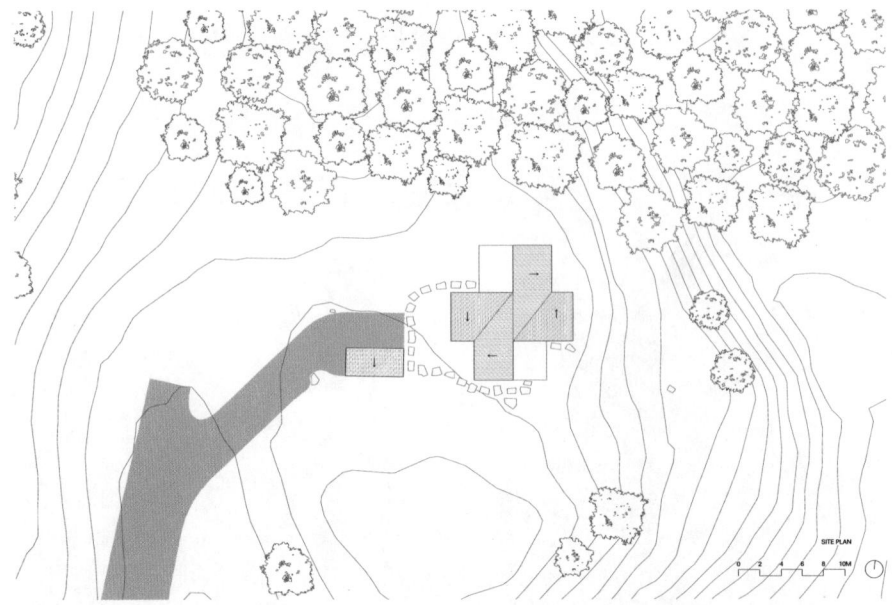

In the case of the Mylla

Concrete also has its own structural, tactile, light transforming – and elemental qualities.

프로젝트들에는 유리와 나무도 포함되어 있다. 콘크리트 또한 그만의 구조적, 촉각적, 빛 변형 및 기본적인 특성이 있다.

BRICK-Q1: Tell us about your favorite project that you used brick in or another architect's work - interior, facade, etc.

A: Kahn's conversation with brick at the Exeter library is an inspiration of how we can honor the material that you use in a building.

A: 루이스 칸(Louis Kahn)이 엑세터 도서관에서 벽돌과 나눈 대화는 건물에서 사용하는 재료를 어떻게 존중할 수 있는지에 대한 영감이다.

236 Brick, Brick! What do you want to be?

BRICK-Q2: What are the strengths and weaknesses of brick?

A: The strength of brick is in its structural capacity – and its most common current weakness is that its load bearing capacity is now often ignored as its usually only used as a cladding material in lieu of structural capabilities.

A: 벽돌의 장점은 구조적 능력이다. 현재 가장 일반적인 단점은 구조적 기능 대신 클래딩 재료로만 사용되는 경우가 대부분이라 벽돌의 지지 용량이 종종 무시된다는 점이다.

Phillips Exeter Academy Library ©Gunnar Klack

TILE-Q1: Tell us about your favourite project that you used tile in or another architect's work - interior, facade, etc.

A: We do not use tile very much other than in wet areas, but if I had to chose a certain type of architecture that used tile in a holistic way then it would be in Moorish architecture. Particularly when used as screening for privacy and modulation of natural light.

A: 우리는 물기가 있는 영역 외에 타일을 많이 사용하지 않지만, 타일을 전체적으로 사용하는 특정 유형의 건축을 골라야 한다면 무어(Moor) 건축을 고르겠다. 특히 사생활 보호 및 자연 채광의 변조를 위한 스크린으로 사용한다면 말이다.

TILE-Q2: What are the strengths and weaknesses of tile?

A: Its strength is its functional capacity to withstand heat and water, but its weakness is that it is often used to dress up questionable architecture.

A: 장점은 열과 물을 견딜 수 있는 기능성이지만, 단점은 의심스러운 건축물을 가장하는 데 종종 사용된다는 점이다.

GLASS-Q1: Tell us about your favourite project that you used glass in or another architect's work - interior, facade, etc.

A: It is hard not to think of Mies – with either the Farnsworth house or the Friedrichstrasse skyscraper project as

A: 미스(Mies)가 생각나지 않기 어렵다. 판스워스 하우스 (Farnsworth House)든 프리드리히스트라세(Friedrichstrasse)

being the most groundbreaking use of glass in the modernist sense.

고층빌딩 프로젝트든 근대주의에서 그는 가장 획기적으로 유리를 사용했다.

GLASS-Q2: What are the strengths and weaknesses of glass?

A: Glass opens up so many opportunities to allow for more open, light filled structures.
A weakness of glass for us can be when we use too much glazing. It is often a temptation to generate too much transparency - taking away the power to frame views and manipulate light.

A: 유리는 더 개방적이고 가벼운 구조물을 지을 수 있는 더 많은 기회를 열어준다. 우리에게 유리의 단점은 유리를 너무 많이 사용하는 경우이다. 너무 많은 투명성을 만들고 싶은 유혹이 흔하며, 이는 전망을 그리고 빛을 조절할 힘을 앗아간다.

WOOD-Q1: Tell us about your favourite project that you used wood in or another architect's work - interior, facade, etc.

A: Sverre Fehn's Villa Schreiner use of wood both as bearing structure and as finish material is a fantastic project in the way it relates to the site, manipulates light, creates texture and life in a building.

A: 스베레 펜(Sverre Fehn)의 빌라 스크레이너(Villa Schreiner)는 지지 구조와 마무리 재료로 목재를 사용하며, 사이트와 연관되고 빛을 조절하고 건물에 질감과 삶을 만드는 방식이 환상적인 프로젝트이다.

WOOD-Q2: What are the strengths and weaknesses of wood?

A: Wood can be economical, sustainable, have a unique appeal to the human senses (tactility, olfactory, visual) and incredibly versatile - its only real limitation is in extremely large structures, when in direct contact with water or under certain fire constraints.

A: 나무는 경제적이고 지속 가능하며 인간의 감각(촉각, 후각, 시각)에 독특한 호소력을 지녔고 아주 다용도이다. 유일한 실질적인 제한은 매우 큰 구조이거나, 물과 직접 접촉하는 경우, 혹은 특정 방화 조건 하이다.

"MATERIAL CHOICE IS AFFECTED BY MANY THINGS."

Villa Schreiner ©Vidariv

240 Brick, Brick! What do you want to be?

murmuro

Who is ...?

João Caldas (Braga, 1981) is the co-founder, with Rita Breda (Estarreja, 1981) of the office murmuro, working from Braga and Porto, in Portugal. They have started their joint path while still at the university, having graduated from DARQ-FCT, University of Coimbra. As Erasmus students, at the NTNU and the Fine Arts Academy in Trondheim, Norway, they had the opportunity to taken part of an interdisciplinary and collaborative program for architecture and art students, pivotal in their education.

Q1: What is material to an architect (or to you)?

A: Material is what we work with, we create shapes and volumes that together create space, but for all of that we do need material. **Each material has its own characteristics and it is important to be aware of those as well as the techniques or constructive logic associated to them.** The plastic possibilities each material allows are important however without forgetting its proper constructive logic.

A: 우리가 함께 일하는 것이 재료이다. 우리는 모양과 부피를 만들고 그 둘은 공간을 만들지만 그러기 위해서는 재료가 필요하다. **각 재료는 고유한 특징이 있으며, 관련 기술이나 건설 방식뿐만 아니라 특성 자체를 인식하는 것이 중요하다.** 각 재료가 허용하는 가소성의 가능성도 중요하나 적절한 건설 방식을 잊지 않는 것 또한 중요하다.

Q2: Tell us about your favourite (or most often used) material and why.

A: We do not hold a favorite material or one that we prefer to work with. Each project, each site, each program, gives us the clues for the material to use. For instance, we made a project for a temporary pavilion that also had to be itinerary, hence we decided to work with wood and textile for an easier and fastest assembling/disassembling operation. For the refurbishing of a co-working space with a very low budget we worked with very common materials, such as corrugated metal plates and fiberglass panels, similar in design, one opaque and the other translucent.
Each project requires a particular approach in order to solve a given problem.

A: 가장 좋아하는 재료나 선호하는 재료는 없다. 각 프로젝트, 사이트, 프로그램에 써야 할 재료에 대한 단서가 있다. 예를 들어, 우리는 이동 가능한 임시 파빌리온 프로젝트를 만드는데 간편하고 빠른 조립을 위해 나무와 직물을 썼다. 예산이 매우 적었던 공동 작업 공간의 리노베이션을 위해서는 파형 금속판과 유리섬유 패널 같은 아주 흔한 재료를 썼다. 그 둘의 디자인은 서로 비슷했는데 하나는 불투명했고 다른 하나는 반투명했다. 각 프로젝트에 주어진 문제를 풀기 위해는 그에 맞는 특별한 접근을 해야 한다.

Q3: When do you decide the material during the design process and what is your criteria? (e.g. budget, client's preference, design concept, climate, etc.)

A: We believe there is not one criteria that overlaps others. As architects our job is to take all the factors into consideration

A: 우리는 다른 기준과 겹쳐지는 단 한 가지 기준은 없다고 믿는다. 건축가로서 우리

murmuro 243

during the design process: financial, programmatic, characteristics of the site, client's expectations... It is with the manipulation of all, guided by our intellect that a project is born. We do not search for novelty or originality per se, as that would be an empty "gesture". We have a critical approach to the mentioned factors in order to reach the most coherent solution.

의 임무는 디자인 과정에서 예산, 프로그램, 사이트의 특징, 의뢰주의 요구 등 모든 요소를 고려하는 것이다. 프로젝트는 건축가의 지성에 의해 인도되는 이 모든 요소의 교묘한 조작을 통해 탄생한다. 우리는 참신함이나 독창성 자체만 찾지 않는다. 그것은 실속 없는 빈 "제스처"일 뿐이다. 우리는 가장 일관된 해결책에 도달하기 위해 언급된 요소에 관해 비판적으로 접근한다.

Q4: What are some architectural projects that inspired you regarding brick, tile, wood and/or glass? And why?

A: We are inspired by several projects and architects. References are fundamental however as architects vocabulary needs time to absorb influences (coming from other architects or even literature, music and other forms of art or even nature) Obviously when talking about brick we remember a trip we made around Finland to visit Alvar Aalto's works and experiments with brick. One building that stood out was Aalto's Muuratsalo Experimental House. We also have some amazing architecture in brick in Portugal.

A: 우리는 다수의 프로젝트와 건축가에게 영감을 받는다. 참고 자료는 필수지만 다른 건축가나, 문학, 음악, 여러 형태의 예술, 혹은 자연에서 오는 건축가의 디자인 어휘는 그 영향력을 흡수하는 데 시간이 걸린다. 벽돌에 관해 얘기하자면 우리는 단연코 알바 알토의 벽돌을 사용한 작품과 실험물을 방문했던 핀란드 여행이 기억난다. 기억에 남았던 건물은 알토의 무라찰로 실험 주택이었다. 포르투갈에도 굉장히 멋진 벽돌 건축이 있다.

Aalto's Muuratsalo Experimental House

Q5: Tell us about the materials you are interested in or want to use in your projects right now.

A: Any material is a white canvas for us. They all have their characteristics and possibilities and is the open possibilities or ways to work with it that interest us.

A: 어떤 재료든지 우리에게는 하얀 캔버스이다. 모두 각자만의 특징과 가능성을 가지고 있고 모두 열린 가능성, 혹은 우리에게 흥미로운 방식으로 일할 기회이다.

BRICK-Q1: Tell us about your favorite project that you used brick in or another architect's work - interior, facade, etc.

A: We have worked with brick on the project for the Plátanos School. We have used brick on the facade as a constructive and plastic solution for a technical necessity.
The client had very specific needs: we could not reduce the outside sports area, they did not want HVAC solutions only

A: 우리는 플루타노스 학교 프로젝트에 벽돌을 사용했다. 기술적인 필요성을 위한 건설적이고 가소성이 좋은 해결책으로 파사드에 벽돌을 썼다. 이 프로젝트의 의뢰주는 아주 구체적인 요구 사항이 있었다. 우리는 외부 스포츠 구역을 축소할 수 없었고, HVAC

murmuro 245

natural ventilation taking advantage of the existent wind corridor, we could not open windows towards the sports area for security reasons, the building had to connect two different pre-existent buildings plus two outside spaces all at different heights, had to be built during the functioning of the school and in two phases. Also, the program demanded we used the facade facing the sports area. That is why we came up with the solution of a complete facade in brick, on a hit and miss design. This solution allowed us to have a facade that would stand the games and balls thrown at it, would allow us to open windows behind to introduce natural light into the secondary program of the school, and allow us to disguise several ventilation grids, the ones allowing to inflate fresh and cold air into the building.

BRICK-Q2: What are the strengths and weaknesses of brick?

A: As every material it has its strengths and weaknesses, but also depends on how we look at it and what we make of it. Brick is normally associated with masonry walls, sturdy and robust walls, however we have designed an entirely perforated wall with bricks. It is also very durable and has very low maintenance. As weakness, we would point out that a brick façade as complex as the one we've designed requires highly skilled handwork, which can easily transform a very competitive solution into an expensive one.

린 벽을 디자인했었다. 벽돌은 또한 내구성이 아주 좋고 관리에 손이 많이 가지 않는다. 단점으로, 우리가 디자인했던 것처럼 복잡한 디자인의 벽돌 파사드는 고도로 숙련된 수세공이 필요하다. 이 점은 아주 경쟁력 있는 디자인을 비싼 디자인으로 쉽게 바꾸기도 한다.

GLASS-Q1: Tell us about your favourite project that you used glass in or another architect's work - interior, facade, etc.

A: We have refurbished an office, actually, transformed three small office spaces into a bigger one. In this project we have used glass as a partition material that would allow us to keep the flow and open feel needed. The main workroom, for about 40 people, has an entire side in glass/window frames for the outside, and the two smaller sides also in glass, facing the kitchen and lounge area on one side and the partners office/meeting room on the other.

A: 우리는 사무실 리노베이션을 했었는데 사실 세 곳의 작은 사무실 공간을 하나의 큰 공간으로 완전히 바꾸는 일이었다. 이 프로젝트에서 우리는 필요한 흐름과 개방된 느낌을 유지하게 해주는 유리를 칸막이로 썼다. 대략 40명 정도의 인원을 위한 메인 워크룸은 바깥을 향하는 한 면이 전부 유리와 창문틀이었고 다른 작은 두 면 역시 유리를 썼다. 작은 유리 중 한 면은 부엌과 리운지를 향했고 다른 한 면은 파트너의 사무실과 회의실을 향했다.

GLASS-Q2: What are the strengths and weaknesses of glass?

"ANY MATERIAL IS A WHITE CANVAS FOR US."

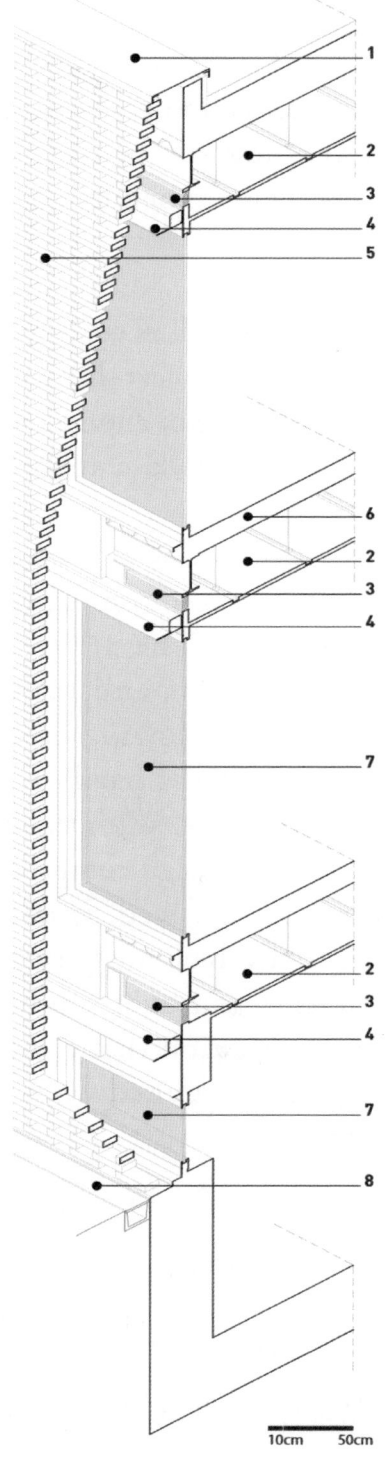

1 zinc sheet flashing
2 ceilling ventilation duct
3 metallic ventilation grid
4 metallic structure for brick
5 permeable brick facade
6 metallic slab
7 exeterior window frames
8 drainage courtyard grid
AXONOMETRIC SECTION

Plátanos School Facade Detail

A: Glass has the capability of defining a physic barrier and at the same time, not. It´s weakness may be that is difficult to achieve acoustic comfort with it.

A: 유리는 정신적 장벽을 정의하는 능력이 있지만 동시에 정의하지 않기도 한다. 유리의 단점은 청각적 편안함을 이루기 어렵다는 점일지도 모른다.

WOOD-Q1: Tell us about your favourite project that you used wood in or another architect's work - interior, facade, etc.

A: We have used wood in a project of an itinerant art pavilion for the Serralves Foundation. The structure, walls, windows, doors and furniture were all in wood, allowing us to assemble / disassemble the pavilion in a simple and fast manner. With this material we were able do design complex connection joints that allowed us to reduce the disassembled building to single elements; a fast, easy and light solution in terms of transportation that also means less storage area required when in between exhibitions.

A: 우리는 세랄브스(Serralves) 재단을 위한 순회 아트 파빌리온에 나무를 사용했다. 구조, 벽, 창문, 문 및 가구는 모두 나무로 되어있어 간단하고 빠르게 파빌리온을 조립 및 분해할 수 있었다. 이 재료로 우리는 분해된 건물을 단일 요소로 줄일 수 있는 복잡한 연결 조인트를 설계할 수 있었다. 운송 측면에서도 빠르고 쉽고 가벼우니 전시회 사이에 필요한 창고 공간도 적었다.

WOOD-Q2: What are the strengths and weaknesses of wood?

A: Wood is an incredible material, it

A: 나무는 상상할 수 있는 거

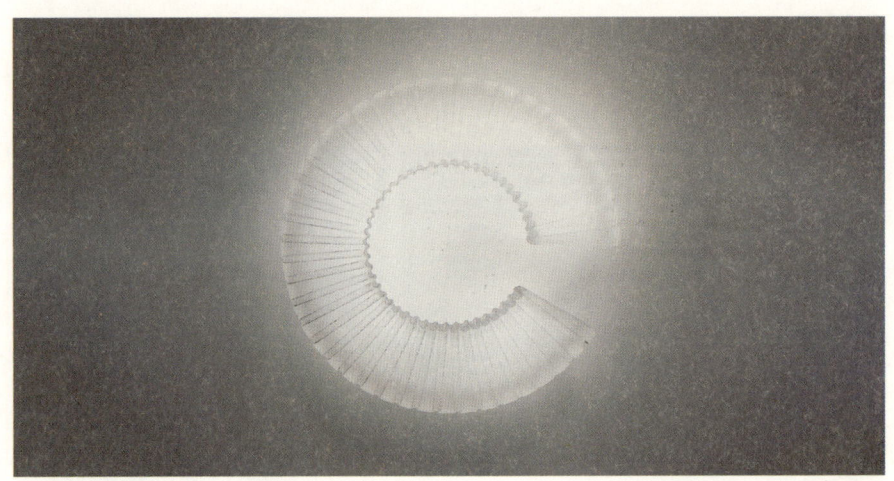

Itinerant art pavilion

can be crafted to almost every detail you can imagine. It gives a sense of warmth and comfort , it connects with people. The biggest weakness of wood is its maintenance as it requires constant care so it doesn't degrade. **If a wood building is well thought and maintained, it can last for centuries.**

의 모든 디테일을 만들 수 있는 아주 굉장한 재료이다. 따뜻하고 편안한 느낌을 주고 사람들과 마음이 통한다. 나무의 가장 큰 약점은 노후화를 줄이기 위해 지속된 관리가 필요한 점이다. **나무 건물은 디자인과 관리가 잘 된다면 수세기 동안 오래갈 수 있다.**

NISHIZAWA ARCHITECTS

Who is ...?

SHUNRI NISHIZAWA was born in 1980. He obtained his B.Arch (2003) and M.Arch (2005) degrees from Tokyo University.

He worked from 2005 to 2009 for Tadao Ando Architect & Associates (Osaka). Then, he worked with Vo Trong Nghia (Ho Chi MInh City, 2009-11).

He was a Partner in Sanuki+Nishizawa Architects (2011-15), before founding Nishizawa Architects in 2015. His work includes the Binh Thanh House(with Vo Trong Nghia Architects, Ho Chi Minh City, 2013); Thong House (Ho Chi Minh City, 2014); Katzden Factory (Binh Duong City, 2016); House in Chau Doc (2017; published here); Pizza 4P's Ben Thanh (Ho Chi Minh City,2017), all the completed buildings are in Vietnam.

Q1: What is material to an architect (or to you)?

A: Material for us is one of important hints for connecting to the building's context.

A: 우리에게 재료는 건물의 콘텍스트에 연결할 중요한 힌트 중 하나이다.

Q2: Tell us about your favourite (or most often used) material and why.

A: Actually we dont have favourite material. It depends on the context of the site. If the site's context have any interesting material, we will make use those one for our design. If not, **the material sometimes will come up with concept design.**

A: 사실 우리는 좋아하는 재료가 없다. 사이트의 콘텍스트에 달려 있기 때문이다. 만약 사이트의 콘텍스트에 흥미로운 재료가 있다면, 그 재료를 디자인에 사용한다. 그렇지 않다면, **재료는 때때로 컨셉 디자인에서 나온다.**

Q3: When do you decide the material during the design process and what is your criteria? (e.g. budget, client's preference, design concept, climate, etc.)

A: We choose the material during the concept phase cause it plays an important role in creating the architecture concept and atmosphere we head for. The criteria for choosing material will depend on what kind of project, the context of the

A: 우리는 콘셉트 단계에서 재료를 선택하는데, 이는 재료가 우리가 추구하는 건축 콘셉트와 분위기를 만드는 데 중요한 역할을 하기 때문이다. 재료의 선택 기준은 사이트 주변의 콘텍스트, 의뢰인의 요구 사항

the National Assembly in Dhaka of Louis Kahn ©Naquib Hossain

surrounding site, client's requirement and their budget, we usually choose the local material of site's region to show off the atmosphere that we want to emphasize for the building and make the original material become more contemporary with the our style design.

Q4: What are some architectural projects that inspired you regarding brick, tile, wood and/or glass? And why?

A: We like the natural expression of the brick itself in the National Assembly in Dhaka of Louis Kahn. Almost of details in this building was made very nice, from the exterior wall to interior wall. And the way he create abstracts form for the facades, the curved wall is also really amazing for us.
About the glass, we like the way SANAA use them in designing almost their building because in every single project of them, we can see a new approach, a new aspect, a new using to the glass that they create. The glass become smoother in curved-shape, harmonize with the concrete structure and the surrounding's context sensitively

및 예산에 따라 달라진다. 우리는 일반적으로 현지 재료를 선택해 건물에 대해 강조하고 싶은 분위기를 과시하고 원래 재료를 우리 스타일로 좀 더 현대적으로 만든다.

A: 루이스 칸(Louis Kahn)이 다카에 있는 국회 건물에 벽돌을 자연스럽게 표현한 것을 좋아한다. 이 건물의 거의 모든 디테일은 외벽에서 내부 벽까지 매우 훌륭하다. 그리고 그가 파사드에 추상적 형태를 만드는 방식과 곡선의 벽은 정말 놀라웠다.
우리는 SANAA가 거의 모든 건물에 유리를 사용하는 방식을 좋아한다. 왜냐하면 모든 프로젝트에서 새로운 접근, 새로운 측면과 그들이 만들어낸 유리의 새로운 사용법을 볼 수 있기 때문이다. 유리는 곡선 모양에서 더 매끄러워지고 콘크리트 구조 및 주변 환경과 민감하게 조화를 이룬다.

Almost of details in this building was made very nice, from the exterior wall to interior wall.

이 건물의 거의 모든 디테일은 외벽에서 내부 벽까지 매우 훌륭하다.

the National Assembly in Dhaka of Louis Kahn ©Asifsaleheen

Q5: Tell us about the materials you are interested in or want to use in your projects right now.

A: Now we are going to use the sunscreen sheets for agriculture because the half-transparent feature have possibilities the create an interesting space under that layer. Beside that, it's also easily to find everywhere cause it's common, cheap, and friendly to everyone in the daily life.

A: 이제 우리는 빛 가림막을 농업에 사용할 것이다. 왜냐하면 반쯤 투명한 특성은 그 층 아래에 흥미로운 공간을 만들 가능성이 있기 때문이다. 게다가 일상 생활에서 흔히 볼 수 있고 가격이 싸며 모든 사람에게 친근한 재료라 어디에서나 쉽게 찾을 수 있다.

BRICK-Q1: Tell us about your favorite project that you used brick in or another architect's work - interior, facade, etc.

A: Katzden Factory is the project we used brick like a main material for the whole building.

A: 카츠덴 공장(Katzden Factory)은 우리가 건물 전체에 주요 재료로 벽돌을 사용한 프로젝트이다.

Katzden Factory Under Contruction ©Hiroyuki Oki

Katzden Factory ©Hiroyuki Oki

BRICK-Q2: What are the strengths and weaknesses of brick?

A: Strength:
+ Help the temperature inside the building become stable, it will be warm in rainy season and cool in hot seasons.
+ Second-hand brick is easy for us to show off the traditional Vietnam colonial architecture's atmosphere under the French domination, combine with new materials make the building look more contemporary than ever.

Weak:
+ Cause the wall need the thickness so we have to use the concrete-beam like a

A: 장점
+ 건물 내부의 온도가 안정되도록 돕는다. 우기에는 따뜻하고 더운 계절에는 시원하다.
+ 중고 벽돌로 프랑스 지배하에 있었던 베트남 식민지 시대의 전통적인 분위기를 보여주기 쉽다. 이를 새로운 재료와 결합하면 건물은 그 어느 때보다 현대적으로 보인다.

단점
+ 벽돌 벽은 두께가 필요하기 때문에 주된 요소마냥 콘

BEN THANH restaurant ©Hiroyuki Oki, NISHIZAWAARCHITECTS

main-factor, brick at the time become like a finishing-layer for exterior wall of the building.

크리트 들보를 사용해야 한다. 그럴 때 벽돌은 건물의 외벽을 위한 마무리 층처럼 되어버린다.

GLASS-Q1: Tell us about your favourite project that you used glass in or another architect's work - interior, facade, etc.

A: BEN THANH restaurant is our favourite project that using glass for the partition inside.

A: 벤 탄 (Ben Thanh) 레스토랑은 내부 칸막이로 유리를 사용한 가장 좋아하는 프로젝트이다.

GLASS-Q2: What are the strengths and weaknesses of glass?

A: Strength
+ Can help to create an in-between space and also connect the space inside and outside together.
+ With its transparent and reflection character, when the sunlight reflect into the glass, the angle of it can help to create the variaty of interesting optical effects with surrounding context at different points of time, weathers or seasons. We think that it's really make sense when that partition can bring a refresh feeling for everyone, everytime they come to our

A: 장점
+ 중간 공간을 만들고 내부와 외부 공간을 연결하는 데 도움이 된다.
+ 투명하고 반사되는 특성은 햇빛이 유리에 반사 될 때 그 각도로 각기 다른 시간, 날씨 또는 계절에 따라 주변 상황에 맞는 흥미롭고 다양한 광학 효과를 만들 수 있다. 칸막이가 모든 사람들에게 건물에 올 때마다 새로운 느낌을 줄 수 있다면 좋겠다고 생각한다.

building design.

Weak
+ Sometimes still need some frames to protect the edges and surfaces. For this case, those glass was like a hand-craft so it's really take time and money to make them reality to our building.
+ Not totally stop the air conditioner escaping from the inside space.
+ Need cleaning up regularly cause it easily get dirty

단점
+ 때로는 가장자리와 표면을 보호하기 위해 프레임이 필요하다. 이 경우, 유리는 수공예품 같아서 건물에 현실로 만드는데 시간과 돈이 많이 소요된다.
+ 내부의 시원한 공기가 빠져 나가는 것을 완전히 막을 수 없다.
+ 쉽게 더러워지기 때문에 정기적으로 청소해야 한다.

WOOD-Q1: Tell us about your favourite project that you used wood in or another architect's work - interior, facade, etc.

A: About our favourite project that use wood in is the Agri Chapel from Yu Momoeda Architects - a Japanese-wooden chapel with a fractal structure system
We love this one because the way they combine the Japanese traditional wood system with the gothic style in contemporary way creatively. And another reason is the way connect the activity of the chapel to the natural surroundings seamlessly.

A: 우리가 가장 좋아하는 나무를 사용한 프로젝트는 모모에다 유우 건축사무소(Yu Momoeda Architecture Office)의 농업 예배당(Agri Chapel)으로, 프랙탈 구조 시스템을 갖춘 일본 목조 예배당이다. 이 프로젝트는 일본 전통 목재 구조 시스템과 고딕 양식을 독창성있게 현대식으로 결합하기에 좋아한다. 또 다른 이유는 예배당의 활동을 원활하게 주변 자연환경과 연결하는 방식이다.

WOOD-Q2: What are the strengths and weaknesses of wood?

A: About the strong point of the wood is that it can be used as structure for the whole building without any concrete. Beside that, the color and texture of natural wood are also the elements that help to show the emotion of the building itself also.

About the weak points, wood for us is also a sensitive material that's difficult to use in different environments so it'll be damaged easily by the temperature, humid, insect infestation. The second is that the structural load is also depended on what kind of wood we use, therefore this structure have to combine with some steel elements for increasing bearing capacity sometimes.

A: 나무의 장점은 콘크리트 없이 건물 전체의 구조로 사용할 수 있다는 점이다. 그 외에도 천연 목재의 색상과 질감은 건물 자체의 감정을 보여주는데 도움이 된다. 우리에게 나무의 단점은 여러 환경에서 사용하기 어려운 민감한 물질이기 때문에 온도, 습기, 곤충 감염으로 쉽게 손상된다는 점이다. 두 번째는 구조적 하중이 사용하는 목재의 종류에 달려 있기 때문에 때로는 지지력을 증가시키기 위해 구조의 일부를 강철과 결합해야 한다는 점이다.

"MATERIAL FOR US IS ONE OF IMPORTANT HINTS FOR CONNECTING TO THE BUILDING'S CONTEXT."

Object-e architecture

Who is ...?

Object-e architecture is an architectural practice currently based in Thessaloniki, Greece and directed by Katerina Tryfonidou and Dimitris Gourdoukis.

It started in 2006, in St. Louis, USA, as a platform with the intention to explore new territories in architecture with the aid of computational tools and techniques. Through time, object-e architecture moved beyond the borders of computation and engaged design at large, trying to graft the new media with the social, political and ecological issues that architecture is facing today. It has won a number of prizes in international competitions, and Its works has been published, exhibited and presented internationally.

Q1: What is material to an architect (or to you)?

A: The materiality of an architectural project is an aspect of the design process that is often overlooked, or, left for the end of that process. In other words, it is quite often considered of secondary importance. That attitude goes in many cases hand in hand with a similar disregard for the architectural detail; the specifics of how a projects is structured and layered. Instead, other aspects are considered more important: form, geometry, function etc. For us, materiality, along with architectural detail, is of equal importance and can define design decision from the beginning of the design process. Materiality holds a vast array of possibilities that expect to be exploited by the architect. That can only be achieved to a full extent only when it is considered as a defining element of the design.

A: 건축 프로젝트의 물질성은 종종 간과되거나 설계 과정의 끝으로 남겨진다. 다시 말해, 물질성은 자주 부차적인 것으로 간주한다. 그런 태도는 많은 경우 건축 디테일에 대한 비슷한 무시와 함께 수반된다. 프로젝트가 어떻게 구조화되고 계층화되는지에 대한 세부 사항이 무시되는 것이다. 대신 형태, 기하학, 함수 등 디자인의 다른 측면이 더 중요하게 여겨진다. 우리에게 있어, 건축 디테일과 함께 물질성은 디자인의 다른 측면과 똑같이 중요하며 설계 과정의 시작부터 디자인을 정할 수 있다. 물질성에는 건축가가 활용할 수 있는 다양한 가능성이 있지만, 설계의 주요 요소로 간주하는 경우에만 전력으로 달성할 수 있다.

Q2: Tell us about your favourite (or most often used) material and why.

A: I don't think that as a practice we have a favorite material. After all, each material has its own use and can be employed

A: 건축 사무소로써 우리가 특히 좋아하는 재료는 없는 것 같다. 결국, 각 재료는 고유한 용도로 사용되며 매번 우리

Object-e architecture 265

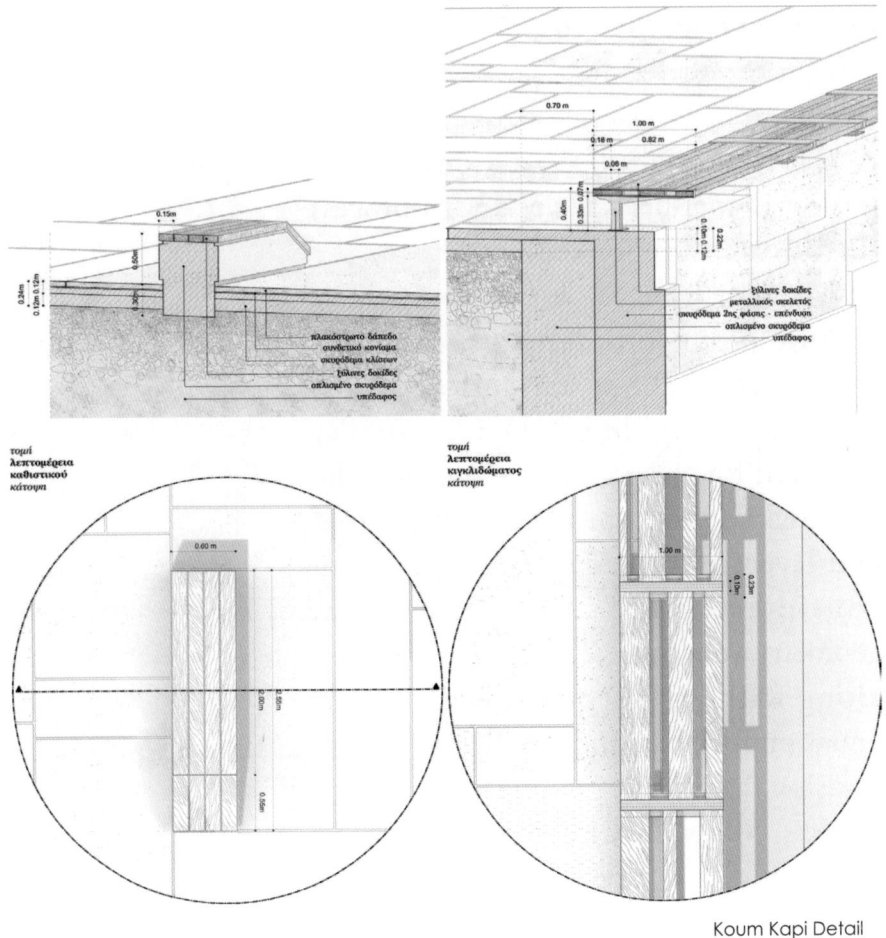

Koum Kapi Detail

when it helps us to create the architectural qualities that we are after each time. That approach towards materials is in line with the general approach of our office which, among other things, is based on the fact that we don't rely on any concept of 'personal style' – where style is understood as a recognizable architectural approach that can create a 'signature'

가 원하는 건축적 특성을 창출 하는데 사용할 수 있다. 재료 에 대한 접근법은 우리 사무 소의 일반적인 디자인 접근법 과 일치한다. 이는 무엇보다도 '개인 스타일'이란 개념에 의 존하지 않는다는 사실에 기반 한다. 여기서 스타일은 건축가 의 특징을 만들어 쉽게 알아볼 수 있는 건축적 접근법이다. 이러한 개인 스타일은 일반적

for the architect. Such a personal style is usually established by architects through the repetitive use of specific formal tropes in several projects and can be enhanced by a consistent use of materials. This tactic might be beneficial in term of becoming recognizable as an architect, but at the same time doesn't help architecture to respond to the conditions that it has to deal with each time. Our approach therefore is different: we understand each project as a unique event. Consequently our projects can be very different to each other; the materials we use and their diversity echo that approach.

While all the above form a fundamental principle in our design process, it doesn't mean that there aren't things that are repeated throughout several of our projects. Each of our projects is a singular event and therefore has its own unique character. But at the same time it incorporates our design principles and our architectural obsessions, which, while also susceptible to change, may create patterns that can be observed in our work.

In that sense we might talk about concrete as a material that reappears in

으로 건축가가 여러 프로젝트에서 주로 특정 형태를 반복적으로 사용하여 만들며, 재료를 일관되게 사용하여 그 스타일을 향상할 수 있다. 이 전략은 건축가로서 이름을 알리는 데에는 유익할 수 있지만, 그와 동시에 건물이 매번 맞서야 하는 조건에 대응하는데 도움이 되지는 않는다. 따라서 우리의 접근 방식은 다르다. 우리는 각 프로젝트를 독특한 이벤트로 본다. 결과적으로 우리의 프로젝트는 서로 매우 다를 수 있다. 우리가 사용하는 재료와 그 다양성은 이런 접근법을 반영한다.

이 모든 것이 우리 설계 과정의 기본 원리를 형성하지만, 그렇다고 해서 여러 프로젝트에서 반복되는재료가 없다는 뜻은 아니다. 우리 프로젝트는 각각 하나의 이벤트이므로 각자의 고유한 특성이 있다. 하지만 동시에 우리의 디자인 원리와 건축적 강박관념도 통합되어 있다. 시간이 지남에 따라 이 강박관념은 변할 수도 있지만, 우리 프로젝트에 일정한 양식을 만들 수도 있다.

그런 의미에서 콘크리트는 우리 프로젝트에서 반복해서 나타나는 재료라고 할 수 있다. 한편으로 콘크리트를 사용하는 것은 실용적인 문제의 결과

our projects. On one hand that is the result of a practical issue: it is a widely used material in construction and can be used to create from standard 'columns and slabs' structures to highly sculptural ones. On the other hand, we like the visual and tactile qualities of exposed concrete and therefore that is another reason we use it.

이다. 콘크리트는 건설에 널리 사용되는 재료이며 '기둥과 슬래브'같은 표준 구조를 매우 조각적인 구조로 만들 수 있다. 또 다른 이유로, 우리는 노출된 콘크리트의 시각적 및 촉각적 특성을 좋아한다.

Cyprus Medical School

Q3: When do you decide the material during the design process and what is your criteria? (e.g. budget, client's preference, design concept, climate, etc.)

A: **The materiality of our projects is a question that we tackle right from the beginning of the design process.** It is not so much a matter of a specific decision that comes from the outside of the process. In other words it is not something that we decide and then we implement. It is a property of the project that gets defined and formulated as we design – in the same way that other aspects of the design do: form, function, geometry, personal intent etc. Materiality is equally important and operates at

A: 프로젝트의 물질성은 설계 과정의 시작부터 바로 다뤄야 하는 문제이며, 과정 외부의 특별한 결정사항이 아니다. 다시 말해, 재료를 결정한 다음 그냥 짓는 것이 아니라는 뜻이다. 물질성은 형태, 기능, 기하학, 개인 의도 등 디자인의 다른 측면과 같이 우리가 설계할 때 정하고 공식화되는 프로젝트의 속성이다. 디자인의 다른 모든 측면과 똑같이 중요하며 동일한 수준에서 동시에 작용한다. 이 모두 일반적으로 변증적이지 않은 방식으로 함께

Robie House © Lykantrop

Object-e architecture 269

the same level and in parallel with all those other aspects. All of them interact together in a usually non-dialectic way and are constantly modulated until the project reaches its final stage. Therefore, the criteria for material decisions lay in the interaction with all other aspects of the design process.

상호 작용하며, 프로젝트가 최종 단계에 도달할 때까지 끊임없이 변조된다. 따라서 재료 결정의 기준은 설계 과정의 다른 모든 측면과의 상호 작용에 있다.

Q4: What are some architectural projects that inspired you regarding brick, tile, wood and/or glass? And why?

A: One of the most remarkable examples of architectural projects using brick is Frank Lloyd Wright's Robie house. While brick is used effectively in order to create solid geometrical volumes, and its color and texture emphasize its connection to the earth, it is maybe the way that it emphasize the horizontal line that is most spectacular in the use of the material and goes several steps further towards expressing the relation with the landscape of the area. Wright uses simple brick in such a way that it almost reads as horizontal lines and not the 'usual' brick pattern. This of course is the result of Wright's attention to detail: for the small, vertical connection between the bricks

A: 벽돌을 사용하는 건축 프로젝트 중 가장 주목할만한 사례는 프랭크 로이드 라이트(Frank Lloyd Wright)의 로비하우스(Robie House)이다. 벽돌은 단단한 기하학적 부피를 생성하는 데 효과적으로 사용되며 색상과 질감은 지구와의 연결을 강조한다. 하지만 여기서 가장 놀라운 점은 수평선을 강조하는 방법이며 주변 풍경과의 관계를 표현하는 데에 있어 몇 단계 더 나아간다. 라이트는 단순한 벽돌을 '보통' 벽돌 패턴이 아니라 거의 수평선으로 읽히도록 썼다. 이는 물론 라이트가 디테일에 주의를 기울인 결과다. 벽돌 사이의 작고 수직적인 연결점에는 붉은 모르타르를 사용하고 더 긴

Robie House © Lykantrop

he uses red mortar, while for the longer horizontal ones white. Therefore the effect of continuous red lines in achieved.

수평의 모르타르에는 흰색을 썼기 때문에 연속적인 빨간 선이 만들어진다.

In terms of glass, Bruno Taut's glass pavilion is another spectacular example of how the possibilities offered by a material can be exploited to the fullest. The glass pavilion, or the Glashaus, was designed and built for the Werkbund exhibition of 1914, in Cologne. Of course, the remaining, black and white photographs of the project reveal only a fraction of the ways in which glass was operating for the project – one can only imagine the effect of the vividly colored glass panels. Still, even without the presence of color the existing representations of the project are enough to reveal the virtuosity with which Bruno Taut worked with the material. From the ethereal, almost unreal feeling of the glass staircase to the sculptural geometry of the glass roof, Taut transcended the use of glass and the properties of the material.

유리의 경우, 브루노 타우트(Bruno Taut)의 유리 파빌리온은 재료가 제공하는 가능성을 어떻게 최대한 활용할 수 있는지 보여주는 또 다른 훌륭한 예이다. 유리 파빌리온 또는 글라스하우스(Glashaus)는 1914년에 쾰른에서 열린 베르크분트(Werkbund) 전시회를 위해 설계되고 지어졌다. 물론 이 프로젝트의 남은 흑백 사진으로는 유리가 어떻게 쓰였는지 그 방법의 아주 일부만을 볼 수 있다. 선명한 색상의 유리 패널이 어떤 효과를 불러일으켰을지 상상만 해볼 뿐이다. 그럼에도 불구하고, 색상이 없어도 사진에서 볼 수 있는 프로젝트는 브루노 타우트가 재료로 달성한 기교를 나타내기에 충분하다. 유리 계단의 미묘하고 거의 비현실적인 느낌에서부터 유리 지붕의 조각적인 기하학에 이르기까지 타우트는 유리의 사용법과 재료의 특성을 초월했다.

TILE-Q1: Tell us about your favourite project that you used tile in or another architect's work - interior, facade, etc.

A: Our project for the redesign of the Koum Kapi waterfront competition (that won the second prize) is to a large

A: 쿰 카피(Koum Kapi) 해안가 재설계 프로젝트 공모전(2등 상 수상)은 거의 타일링 시스템에 대한 연구이다. 우리가

Bruno Taut's glass pavilion

Object-e architecture

extend a study of a tiling system. Aim of the proposed intervention is the design of a public space that will be able to address both the local (at the scale of the neighborhood) and the city level. The proposal is organized through the subdivision of the ground plane, that follows an approach that departs from what we usually expect in similar situations, that is, linear tracings along the seaside that are defining zones of different movements and activities. Instead the ground plane is subdivided in rectangular shapes of different sizes and materiality. Therefore the design is structured around different ways according which tiles can be arranged and manipulated in order

제시한 디자인의 목표는 지역(동네 규모)과 도시 모두를 맡을 수 있는 공공 공간의 설계이다. 이 프로젝트는 지상 평면의 세분하여 구성된다. 여기서 우리는 유사한 상황에서 일반적으로 기대하는 것, 즉 해변을 따라 각기 다른 움직임과 활동의 영역을 정하는 접근법을 피했다. 대신 지상 평면은 크기와 물질성이 다른 직사각형 모양으로 세분된다. 따라서 경로를 형성하고, 이동과 방향을 정하고, 기능을 차별화하고, 연속성을 만들고, 중요한 영역을 나타내기 위해 타일을 배열하고 조작할 수 있는 다양한 방식으로 디자인이 구성된다.

Koum Kapi Project

to form paths, define movement and direction, differentiate functions, create continuities and signify important areas. The subdivision and tiling is controlled through a packing algorithm. The parameters of the algorithm helped us control density, direction and position of the tiles. The maximum and minimum size of the packing process is altered locally according to the different events that each area will host. The result is a patchwork of sizes and materials, that apart from its main organizational properties it also functions as a reference to both the nature of the Koum Kapi area as a patchwork of diverse elements and to patterns that can be found throughout the city of Chania (for example the masonry of the city's walls). The size of the rectangulars range from a few centimeters to 10 meters, allowing this way for a large variation.

Along this patchwork various 'spacial events', of larger or smaller scale, are taking place that are usually signified with the presence of the larger rectangular shapes. The tiling functions more like a field, whose elements are of varying and locally differentiated size and density. Therefore the result of the proposal

세분화와 타일 공사는 포장 알고리즘으로 조절했다. 알고리즘의 변수는 타일의 밀도, 방향 및 위치를 제어하는 데 도움이 되었다. 포장 프로세스의 최대 및 최소 크기는 각 영역이 개최할 여러 이벤트에 따라 국소로 변경했다. 그 결과는 다양한 크기와 재료의 패치워크이며, 주요 구조적 속성과는 별로도 코움 카피 지역의 다양한 요소의 패치워크 같은 자연과 차니아(Chania)시 전역에서 발견할 수 있는 패턴을 나타낸다 (예: 도시 벽의 석조). 타일 직사각형의 크기는 몇 cm에서 10m까지이며, 덕분에 타일 패턴에 폭 넓은 변형이 가능하다.

이 패치워크를 따라 더 크거나 작은 규모의 다양한 '공간 이벤트'가 발생하며, 일반적으로 더 큰 직사각형이 있음을 뜻한다. 타일 패턴은 요소가 다양하고 국부적으로 차별화 된 크기와 밀도를 가진 들판과 비슷하다. 따라서 디자인의 결과는 동시에 디자인의 국부적인 차별화를 촉진하고 해변 전체에 눈에 띄는 어휘를 만드는 도시의 들판이다.

is an urban field that at the same time promotes local differentiation and provides a recognizable vocabulary throughout the seaside.

Our project for the Souda Ferry Terminal competition (honorable mention), aims to create a unique, narrative experience for the user by challenging what we expect to encounter in buildings with such 'heavy' uses, which need to accommodate large numbers of visitors in transit. The detailing and character of the proposal, that wouldn't be out of place in very different architectural typologies, references a time when travel for leisure was a form of luxury and therefore the buildings that were meant to support it were echoing that character. At the same time, the simple functional layout of the building and the materials used assure that the proposed design will be able to sustain the very high load of use that results from the number of visitors of a ferry terminal. However, the building tries to transform a place that you are supposed to move through in order to go somewhere else into one that you could probably want to be – by affirming the contradictions that may arise from such a reversal.

수다 페리 부두(Souda Ferry Terminal) 공모전(장려상)을 위한 우리 프로젝트는 많은 방문객을 수용해야 하는 유동인구가 많은 건물에 보통 예상하는 것과 다르게 방문객에게 독특하고 서술적인 경험을 주는 것을 목표로 한다. 매우 다른 건축 유형에서는 그다지 특별하지 않은 이 디자인의 디테일과 특성은 여가를 위한 여행이 사치였던 시대를 나타내기 때문에 그를 위한 건물도 그 특성을 반영한다. 동시에 건물의 간단하고 기능적인 레이아웃과 재료는 이 디자인이 페리 터미널 방문자 수에서 오는 매우 높은 사용량을 지탱할 수 있음을 보장한다. 하지만 이 건물은 사람들이 다른 곳으로 이동하기 위해 통과해야 할 장소를 머물고 싶은 곳으로 바꾸려고 하고 이러한 반전으로 생길 수 있는 모순을 긍정적으로 해결하려 한다.

The ways in which we employ tiles therefore echoes the above: Colored concrete tiles are employed both on the interior and exterior of the station in order to provide a feeling of 'luxury' to a place that is connected to mass tourism. We are trying to imagine our proposal as a building that will recreate the feeling of the station as a place where something special is happening; where slowness can maybe take the place, if only as a feeling and not an actual condition, of the speed that we usually encounter in such places. Using concrete tiles, along with the other material selections, enhance that feeling.

따라서 우리가 타일을 사용하는 방식은 위의 내용을 반영한다. 색이 있는 콘크리트 타일은 대중적이고 관광과 연결된 장소에 '고급스러운 느낌'을 주기 위해 역의 내부와 외부에 사용된다. 우리는 이 프로젝트를 특별한 일이 일어나는 장소였던 역의 느낌을 재현할 건물로 상상한다. 실제가 아닌 느낌만으로라도 이런 장소에서 볼 수 있는 일반적인 속도를 대신하여 느림이 있을 수 있는 곳으로 본다. 다른 재료와 함께 콘크리트 타일을 사용하여 그 느낌을 향상한다.

Souda Ferry Terminal Project

Object-e architecture

TILE-Q2: What are the strengths and weaknesses of tile?

A: **We tend to think of materials in architecture not in terms of strengths or weaknesses, but rather of possibilities.** Therefore we are interested in the possibilities that tiles can offer to us. Tiles of course are based on the idea of repetition. An element has to be repeated, one way or another, in order for the tiling system to be created. Therefore we find a lot of possibilities in the rules that we can apply to the repetition of the tiles. By incorporating difference in the repetition process, many unique condition can arise. Difference in the case of the tiles can be related to position, size and material properties/color; or – maybe more interestingly – to a combination of the above. Through such techniques tiles can stop being a single element on a grid that can create a surface and become many more.

A: **우리는 건축에서 재료를 강점이나 약점의 관점에서가 아니라 가능성의 관점에서 생각하는 경향이 있다.** 그러므로 우리는 타일이 우리에게 제공할 가능성에 관심이 있다. 물론 타일은 반복이라는 개념에 기초를 두고 있다. 타일 시스템을 만들려면 한 요소를 어떠한 방식으로든 반복해야 한다. 그래서 우리는 타일의 반복에 적용할 수 있는 규칙에서 많은 가능성을 찾는다. 반복 과정에 특색을 포함하여 독특한 상태가 많이 발생할 수 있다. 여러 타일 사이의 차이점은 위치, 크기 및 재료 특성/색상과 관련될 수 있다. 또는 더 흥미롭게, 위의 조합에서 나올 수도 있다. 이러한 기술을 통해 타일은 그리드의 단일 요소가 아니라 표면을 생성하고 더 많은 것이 될 수 있다.

GLASS-Q1: Tell us about your favourite project that you used glass in or another architect's work - interior, facade, etc.

A: Our project for the Cyprus Medical School competition is based on the principle that a medical school, as any other university building, shouldn't be just a place where the students go in order to attend classes and then leave. Instead the proposal tries to imagine the design proposal as a place where the students, along with instructors and administration stuff, spend an important part of their day and should therefore offer them high quality spaces not only for their main activity, but also in the spaces intended for leisure, meeting and rest. In that direction the proposal echoes a cloister organization: It expands to all four edges of the site allowing for the creation of a large open space in the center of the building. This space, organized in levels and planted extensively, is understood as the heart of the building. That interior courtyard, while large in size and the main organizational element of the building, remains hidden from the outside. Aim was to function as a secret space; a surprise for the visitor that approaches the building. The exterior of the building on the other hand becomes 'monolithic': it leaves only a couple of small openings through which you can see the interior courtyard while some vegetation that

A: 키프로스 의과대학 공모전 프로젝트는 다른 대학 건물처럼 학생들이 수업에 참석하고 바로 떠나기 위해 가는 곳이 되어서는 안 된다는 신조로 디자인했다. 대신에 학생들은 물론 강사와 행정 직원까지 하루의 중요한 부분을 소비하는 곳이므로 주요 활동뿐만 아니라 여가, 회의 및 휴식을 위한 공간에서도 고품질 공간을 제공하는 장소로 생각했다. 그런 면에서 이 디자인은 회랑의 구성을 되풀이한다. 건물의 중심에 큰 열린 공간을 만들 수 있도록 건물을 사이트의 네 가장자리까지 확장했다. 층으로 구성되고 광범위하게 식물이 심어진 이 공간이 건물의 핵심이다. 안뜰은 크기가 크고 건물의 주요 구성 요소이지만 외부에서는 보이지 않는다. 디자인의 목표는 이 안뜰이 건물에 접근하는 방문객에게 뜻밖의 발견이 되도록 비밀 공간이 남는 것이다. 반면에 건물의 외관은 '단일체'이다. 내부의 안뜰을 볼 수 있는 몇 개의 작은 구멍만 남고 그 입면에 나타나는 일부 식물이 내부에 무엇이 존재하는지에 대한 힌트를 제공한다.

Cyprus Medical School Project

appears on the elevations provides some hints for what exists on the inside.

Classrooms, labs and offices are placed on the outside perimeter of the building, while the interior is occupied by a large corridor facing the courtyard through transparent glass panels. The interior elevations therefore – the sides of the building that face the atrium – are transparent: A glass elevation opens from the building towards the courtyard allowing a constant visual connection between the atrium and the inside of the building. The glass is structured in a rhythmical way though mullions that vary, generating a non-repetitive pattern. The transparency of the glass is controlled only through vegetation that rises in front of the glass panels in places that full transparency in not required.

교실, 실험실 및 사무실은 건물의 바깥 둘레에 배치했고 내부는 투명 유리 패널을 통해 안뜰을 마주 보고 있는 큰 복도로 점령된다. 따라서 아트리움을 마주 보고 있는 건물 측면의 내부 입면은 투명하다. 건물에서 안뜰 쪽으로 유리로 된 입면이 열리며 아트리움과 건물 내부를 일정하게 시각적으로 연결한다. 유리는 리드미컬한 방식으로 구성되지만 다양한 멀리온(mullions)으로 반복되지 않는 패턴을 만든다. 유리의 투명도는 충분한 투명도가 요구 되지 않는 곳에서 자라나는 식물을 통해서만 조절된다.

GLASS-Q2: What are the strengths and weaknesses of glass?

A: "The longing for purity and clarity, for glowing lightness and crystalline exactness, for immaterial lightness and infinite liveliness found a means of its fulfillment in glass—the most ineffable,

A: "순도와 선명도, 빛나는 가벼움과 투명한 정확성에 대한 열망, 물질적이지 않은 가벼움과 무한한 활기에 대한 갈망은 유리에서 이를 성취할 수단을 발견했다. 가장 형언할 수 없

most elementary, most flexible and most changeable of materials, richest in meaning and inspiration, fusing with the world like no other. This least fixed of materials transforms itself with every change of atmosphere. It is infinitely rich in relations, mirroring what is above, below, and what is below, above. It is animated, full of spirit and alive [...]"
Adolf Behne

As with tiles, we tend to think of glass in terms of the possibilities that it might offer. Adolf Behne's quote summarizes those possibilities of glass as a material in the best possible way: transparency and reflections. Transformation of the material that echo changes in the atmosphere, the weather or the season. An almost immaterial quality that can be found between absence and presence. Maybe one of the main properties of glass is exactly this: it can occupy – with the proper design decisions – any place on the continuum that connects some of the most obvious architectural oppositions: transparency and opacity, presence and absence, visibility and obstruction of vision; glass can acquire virtually all possible points in between those extremes and therefore generate very complex conditions.

고, 가장 기본적이고, 가장 유연하고, 가장 변화할 수 있는 재료로, 의미와 영감이 가장 풍부하며 그 어떤 재료와도 다르게도 주변 세계와 융합된다. 이 가장 불변의 재료는 대기의 모든 변화에 따라 변한다. 위에 있는 것을 아래로, 아래에 있는 것을 위로 반사하여 관계가 무한히 풍부하다. 유리는 생기 있고, 활기차며 살아있다 [...] "
아돌프 베네

타일과 마찬가지로 우리는 유리가 제공할 가능성의 관점에서 유리를 생각하는 경향이 있다. 아돌프 베네의 인용문은 유리의 가능성을 가능한 가장 좋은 방법, 즉 투명성과 반영으로 요약한다. 재료가 대기, 날씨 또는 계절의 변화를 반영하여 변형하고 부재와 존재 사이에서 발견할 수 있는 거의 비물질적인 특성을 지닌 것이다. 유리의 주요 특성 중 하나는 정확히 다음과 같다. 투명도와 불투명도, 존재 및 부재, 시야 및 시야 방해와 같은 가장 명백한 반대 요소를 연결하는 연속체의 어느 곳이든 차지할 수 있다는 점이다. 유리는 이러한 극단 사이의 거의 모든 가능한 점을 획득할 수 있으므로 매우 복잡한 조건을 만든다.

OFIS arhitekti

Who is ...?

Based in Ljubljana, formed by Rok Oman and Spela Videcnik (1998)

Rok Oman
(born 1970) studied architecture at Ljubljana School of Architecture (grad.1998) and at Architectural Association in London(grad.2000).
Currently teaching in Harvard GSD, Boston, MA

Spela Videcnik
(born 1971) studied architecture at Ljubljana School of Architecture (grad.1997) and at Architectural Association in London (grad.2000).
Currently teaching in Harvard GSD, Boston, MA

Q1: What is material to an architect (or to you)?

A: **It is the physical definition of the potentialities and limitations of architecture.** We find particularly interesting the use and combination of diverse materials to explore plastic expressions and to develop different patterns and assemblies that reinforce the architectural concept behind.

A: 재료는 건축의 잠재력과 한계에 대한 물리적 정의이다. 우리는 특히 창조적 표현을 탐구하고 건축 콘셉트를 강화하는 여러 패턴과 집합을 개발하기 위한 다양한 재료의 사용과 조합에 흥미가 있다.

Q2: Tell us about your favourite (or most often used) material and why.

A: We are really fond of wood, particularly for its quality to qualify interiors. **Together with wood, they create a powerful combination.**

A: 우리는 목재를 정말 좋아한다. 특히 실내를 특성화하는 속성이 마음에 든다. **나무와 실내는 강력한 조합이다.**

Q3: When do you decide the material during the design process and what is your criteria? (e.g. budget, client's preference, design concept, climate, etc.)

A: It is something that arises at some point, but usually not at the beginning. As it depends on several limitations, such as costs estimation, environment,

A: 재료는 디자인 과정 중 떠오르는 것이지만, 일반적으로 처음에 나타나지 않는다. 비용, 환경, 유형, 정체성과 같은 여러 제한에 따라 다르기 때문

typology, identity, we usually incorporate it during the process and can be adjusted until last minute if necessary.

Q4: What are some architectural projects that inspired you regarding brick, tile, wood and/or glass? And why?

A: Vernacular architecture is always an inspiration for us, nevertheless, not exclusively.

Q5: Tell us about the materials you are interested in or want to use in your projects right now.

A: We are open to experimentation.

GLASS-Q1: Tell us about your favourite project that you used glass in or another architect's work - interior, facade, etc.

A: So far, the most interesting project we have used glass for is the Glass House in the desert in Spain. The unit uses the vertical glazing panels of the envelope as structural walls, resisting the desert's high-speed winds and supporting the timber stressed skin roof and deck. The glass thermally efficient envelope is constructed of triple glazing walls that, due to use of almost invisible coatings, protects the interior from the sun.

외피로 된 지붕과 단을 받친다. 열 성능이 효율적인 유리 외피는 삼중 유리 벽으로 구성되어 있으며 거의 보이지 않는 코팅을 사용하여 태양으로부터 내부를 보호한다.

GLASS-Q2: What are the strengths and weaknesses of glass?

A: One strength is the diversity of natures or physical presences it can acquire. Fabricating a glass element is a very complex process, but at the same time it is possible to create tailored elements that can suit an enormous variety of environments, conditions and atmospheres.
Something to comes to our mind as a weakness of glass is the necessity to use auxiliary materials to install it. There are always limitations on the need to use profiles, substructures, limit the sizes of the panels, create joints, etc.

A: 한 가지 장점은 유리로 얻을 수 있는 특성이나 물리적 존재감이 다양하다는 점이다. 유리 요소를 제작하는 것은 매우 복잡한 과정이지만 동시에 엄청나게 다양한 환경, 조건 및 대기에 적합한 맞춤형 요소를 만들 수 있다.
유리의 단점으로서 생각나는 것은 설치를 위해서 유리 외에 추가로 재료를 사용해야 한다는 점이다. 종단면, 하부 구조, 패널 크기 제한, 이음매 등 때문에 항상 제한이 있다.

Glass House Construction Process ©OFIS

Glass House ©JOSE NAVARRETE

OFIS arhitekti 287

OOIIO Architecture

Who is ...?

OOIIO is an international team of architects, designers and engineers engaged in finding this special "I don´t know what it is" that makes a work unique, exciting and able for transmit sensations on a way that a vulgar work will never get.

Q1: What is material to an architect (or to you)?

A: A material for me is like a word for a writer or like a color for a painter. Is the tool that you use to express your architecture, to solve technically a problem, to keep a project on budget.

A: 내게 재료는 작가에게 단어나 화가에게 색과 같다. 건축을 표현하고, 기술적으로 문제를 해결하고, 예산을 책정하는 데 사용하는 도구이다.

Q2: Tell us about your favourite (or most often used) material and why.

A: As architect I don't have a favorite material, It would be extremely complicated for me to choose one and, even more difficult, to reject all the other materials as "non favorites" for me. Each project is asking to emerge one specific material over the others. The project chooses the right material, not me.
Our projects have always different solutions for each problem, with the most appropriate material each time.
It is true that because of the type of architecture I like to do, at the end of the day I mostly use traditional materials worked in a contemporary way, more than high tech materials and details.

A: 건축가로서 나는 가장 좋아하는 재료가 없기 때문에, 한 재료를 꼽는 것이 어렵다. 그보다 더 어려운 것은 다른 재료 모두를 "좋아하지 않는" 것으로 치부하는 일이다. 프로젝트마다 많은 재료 중 한 가지 특정한 재료가 명백해지기 마련이다. 내가 아니라 프로젝트가 적절한 재료를 선택한다. 우리 프로젝트는 항상 각 문제에 대해 다 다른 해결책이 있고 매번 가장 적절한 재료를 사용한다.
내가 좋아하는 건축 양식 때문에 결국 첨단 기술로 된 재료와 디테일보다는 주로 전통적인 재료를 현대적으로 사용하는 것이 사실이다.

Q3: When do you decide the material during the design process and what is your criteria? (e.g. budget, client's preference, design concept, climate, etc.)

A: It depends, in each project the right material comes up on a different moment. It is always decision coming from a mix of factors like design concept, local tradition with materials, site regulations, better sustainable behavior with local weather, etc.
The solution sometimes comes at the very beginning of the design process and sometimes at the end.

A: 프로젝트마다 적절한 재료는 다 다른 순간에 나타난다. 항상 디자인 콘셉트, 지역 전통의 재료, 사이트 규정, 지역 날씨에 알맞은 지속 가능한 디자인 등과 같은 여러 요인이 혼합되어 결정한다. 해결책은 때로 설계 프로세스의 시작 부분에서, 때로는 끝 부분에서 나온다.

Q4: What are some architectural projects that inspired you regarding brick, tile, wood and/or glass? And why?

A: Brick-any of the ancient buildings following "mudejar" tradition of brick use in Spain.
"Mudejar" is the name given to Muslims of Spain who remained in the country after the Christian Reconquista but were not initially forcibly converted to Christianity (XV th century).
By extension, "Mudéjar" refers to an architecture and decoration style in

A: 벽돌 스페인에 있는 "무데하르(mudejar)" 전통식으로 벽돌은 쓴 고대 건물 중 어느 것이든 좋다. "무데하르"는 기독교 레콩키스타(Reconquista) 이후 처음에는 강제로 기독교로 개종하지 않아도 되었던 스페인에 남은 무슬림에게 주어진 이름이다 (15세기). 그 연장으로, "무데하르"는 무어(Moor) 문화의 취향과 솜씨에 크게 영향을 받

(post-Moorish) Christian Spain that was strongly influenced by Moorish taste and workmanship, reaching its greatest expression in Medieval Spain.
I love it! Is like they draw with bricks!

은 기독교 스페인의 건축과 장식 스타일(포스트-무어)을 말하며 스페인 중세시대에 그 절정을 이르렀다.
나는 정말 좋아한다! 마치 벽돌로 그림을 그리는 것 같지 않은가!

Mudéjar ©José Luis Filpo Cabana

Glass-"Harpa", Iceland Concert and Opera House in Reykjavík, Iceland, with a fantastic colored glass facade inspired by the basalt landscape of Iceland, designed by the Danish firm Henning Larsen Architects in co-operation with Danish-Icelandic artist Olafur Eliasson.

It is an impressive well executed singular glass façade to be seen from exterior and also from the interior of the building. Good material choose for a cold and dark weather like the Icelandic one.

Tile-Chapel of "Las Almas" in Porto, Portugal (or many other churches and palaces in Portugal), with their awesome hand drawn tiles showing any kind of religious scenes, or battles, or animals, or

유리
아이슬란드 레이캬비크 (Reykjavík)에 있는 아이슬란드 콘서트 및 오페라 하우스인 "하르파(Harpa)"이다. 아이슬란드의 현무암으로 된 경관에서 영감에서 나온 환상적인 색을 지닌 유리 파사드가 있다. 덴마크 회사 헤닝 라슨 건축가(Henning Larsen Architects)가 덴마크/아이슬란드 예술가 올라퍼 엘리아슨 (Olafur Eliasson)과 협력하여 디자인했다.
건물 외부는 물론 내부에서도 보이도록 디자인된 놀랍도록 잘 지은 단일 유리 파사드이다. 유리는 아이슬란드처럼 춥고 어두운 날씨에 적합한 재료이다.

타일
포르투갈의 포르토 (또는 포르투갈의 다른 많은 교회와 궁전)에 있는 "라스 알마스"(Las Almas – 영혼) 예배당이 좋다. 모든 종류의 종교적 장면이나 전투, 동물 또는 범선 등이 손으로 그려진 멋진 타일이 있다. 개인적으로 이 전통 공예의 가소성과 현대 건축에서의 잠재성이 굉장히 인상적이고 매력적이다.

Harpa ©Evgeniy Metyolkin

sailboats, etc.
It is absolutely impressive and attractive for me the plasticity and potential use of this craft tradition in nowadays architecture.

Wood-Final Wooden House by Sou Fujimoto. I fall in love of this project from the first time I saw a picture of it.
I have studied it, its details, the strong conceptualization and radical idea behind its architecture, the light how it gets inside.
I would love to visit it one day!

나무 후지모토 소우의 마지막 나무 주택 (Final Wooden House). 이 프로젝트를 처음 본 순간부터 사랑에 빠졌다. 나는 이 프로젝트의 디테일, 건축의 배경에 있는 강한 컨셉과 급진적인 아이디어, 그리고 빛이 내부로 어떻게 들어오는지 연구했었다. 언젠가 꼭 방문해보고 싶다!

Wooden House ©Kenta Mabuchi

Q5: Tell us about the materials you are interested in or want to use in your projects right now.

A: Craft materials. Traditional techniques. Expressive materials with textures, color possibilities and variations able to bring happiness and charm to the projects. Cheap and common materials reinterpreted in new and original ways in architecture. Local materials that are related with local culture and local constructive tradition.
Totally out of high tech cold and non-sustainable solutions.

A: 수제 재료. 전통적 기술. 질감, 색상 및 변형의 가능성이 넘치고 표현력이 다채로운 재료는 프로젝트에 행복과 매력을 가져올 수 있다. 싸고 일반적인 재료를 건축에서 새롭고 독창적인 방식으로 재해석하는 일. 지역 문화 및 지역 건설 전통과 관련된 지역 재료. 첨단의 냉정하고 지속 불가능한 방식은 완전히 안중에 없다.

BRICK-Q1: Tell us about your favorite project that you used brick in or another architect's work - interior, facade, etc.

A: Probably the project where I have used brick with more success is in the refurbishment of an ancient Haystack that was abandoned and falling apart in a village in central Spain.
There were some poor ancient construction used in old times to keep agricultural tools and animals to work in the countryside.

A: 아마도 내가 벽돌을 좀 더 성공적으로 사용했던 프로젝트는 스페인 중부의 한 마을에서 버려지고 붕괴하고 있는 오래된 헛간의 보수 공사일 것이다. 옛날에 시골에서 일할 때 필요한 농업 도구와 동물을 위해 사용된 오래된 건축물이었다. 오래된 벽 중 하나는 아름다운 옛날식 큰 벽돌로 되어있었다. 이 벽이 무너지고 있어서 의뢰

OOIIO Architecture 295

One of the old walls was made from beautiful ancient big bricks and the client wanted to demolish it because it was falling down. We convinced him to keep it and thanks to this special interest in recovering this brick wall we decided to work always with brick in all the new elements introduced in the buildings, with different patterns.

The result of all this architectonic surgery work has been a patchwork of nuances and details rethinking the old and mixing it with the new, turning the result into a mosaic of nuances and details.

Brick here was definitely the right material to merge old with new!

인은 철거하고 싶어했지만, 우리는 벽을 보존하는 쪽으로 설득했다. 이 벽돌 벽을 복구하는 데 특별히 흥미가 있었기 때문에 건물에 도입하는 모든 새로운 요소 역시 전부 벽돌로 하기로 했다. 이 모든 건축적 수술 작업의 결과는 오래된 것을 다시 생각하고 새로운 것과 혼합하여 낳은 뉘앙스와 디테일의 패치워크였다. 여기서 벽돌은 오래된 것과 새로운 것을 결합하는 데 확실히 적합한 재료였다!

Former Haystack Refurbishment

BRICK-Q2: What are the strengths and weaknesses of brick?

A: Strengths: low maintenance, cheap material (in Spain), easy to work with, can solve structure and façade at the same time.
Weaknesses: not the best climatic behavior with sun heat, it works as a heat accumulator.

A: 장점-낮은 유지 보수, (스페인에서는) 저렴함, 다루기 쉬운 점, 구조와 외관을 동시에 해결할 수 있는 점. 단점-태양열에 알맞은 열적 거동이 없고 열을 축적하는 점.

TILE-Q1: Tell us about your favourite project that you used tile in or another architect's work - interior, facade, etc.

A: Definitely the refurbishment of two squares in Talavera de la Reina, a 75,000 inhabitant town in Spain.
I decided to use local Talavera's traditional pottery as the main material, as it is a product of excellent plastic qualities and deeply rooted in local culture. Usually it used with drawings and motifs from the Renaissance in small-scale domestic objects. I propose greatly increase the size of those drawings that adorn vases and dishes up to the urban scale, showing new possibilities for this fabulous craft material.

A: 단연코 7만5천명의 주민이 사는 스페인의 탈라베라 델 라 레이나(Talavera de la Reina) 시에 있는 광장 두 개의 보수 공사이다. 여기서 나는 현지 탈라베라 전통 도자기를 주요 재료로 사용하기로 했다. 타일은 훌륭한 가소성을 지닌 제품이며 지역 문화에 깊이 뿌리를 두고 있기 때문이다. 일반적으로 타일은 르네상스 시대의 그림과 모티프로 소규모 가정적 물체에 쓰이지만 나는 화병과 접시를 장식하는 그림을 도시 규모로 크기를 크게 늘려 이 멋진 공예품에 대한 새로운 가능성을 제안했다.

OOIIO Architecture 297

TILE-Q2: What are the strengths and weaknesses of tile?

A: Strengths: amazing plastic possibilities, good to keep rain water out of your walls, in some places like the city were we work with it, there is a deep tradition with it and great craft artisans.
Weaknesses: You need to glue it really well on the walls to avoid tiles jumping out.

A: 장점-놀라운 가소성, 빗물을 잘 막는 점, 우리가 일하는 도시에서와 같이 일부 장소에서는 깊은 전통과 훌륭한 장인이 있다는 점.
단점-타일이 튀어 나오지 않도록 벽에 잘 붙일 필요가 있다는 점.

ELEGIMOS LOS MOTIVOS INSPIRADOS EN LA TRADICION

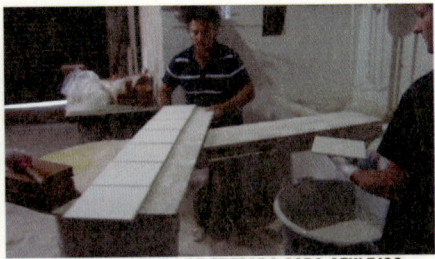
ARTESANALMENTE SE PREPARA CADA AZULEJOS

LOS ARTESANOS LOCALES LLENAN DE MAGIA LAS PIEZAS

POCO A POCO EL COLOR INUNDA LOS MOTIVOS

ORGANIZANDO CADA DETALLE PARA LLEVAR A OBRA

SE TRANSPORTA EN CAJAS CON MUCHO CUIDADO

GLASS-Q1: Tell us about your favourite project that you used glass in or another architect's work - interior, facade, etc.

A: For us, glass never has been the main material of any of our projects. Of course we used glass in every building that we did, in windows, dividing walls, etc. but for the moment never as the main

A: 유리는 우리 프로젝트의 주요 재료가 된 적이 한 번도 없다. 물론 우리가 한 모든 건물에서 창문, 칸막이 등에 유리를 사용해왔다.

PERFILANDO CADA PIEZA

SE ORGANIZAN LOS LIENZOS MEDIANTE CODIGOS

AL HORNO PARA LACAR

COMPROBANDO QUE TODAS LAS PIEZAS ESTAN BIEN

RECEPCION EN OBRA

PIEZA A PIEZA SE VAN LEVANTANDO LOS MOSAICOS

Hand Craft Process of Tiles mosic creation

material that define a project. I am sure it is because until now I have been working on regions where the sun heat is really strong especially in summertime, like central Spain, were we have our office.

Glass is not the most appropriate material over there. Also the local architectural tradition is based more in opaque and massive materials than in light clear ones like glass.

하지만 지금은 프로젝트를 정의하는 주요 재료가 아니다. 나는 이게 지금까지 우리 사무실이 있는 스페인 중부지방 같이 여름철에 특히 태양열이 굉장히 강한 곳에서 일해왔기 때문이라고 확신한다. 그런 지역에서 유리는 적절한 재료가 아니다. 또한 지역의 전통적인 건축은 유리와 같이 가볍고 맑은 것보다 불투명하고 무거운 재료에 더 기반을 두고 있다.

"A MATERIAL IS LIKE A WORD FOR A WRITER OR LIKE A COLOR FOR A PAINTER."

PLAZA DEL SALVADOR REHABILITATION

GLASS-Q2: What are the strengths and weaknesses of glass?

A: Strengths: There is no material like glass, it is a must for most of constructions to solve interior-exterior situations. It is 100 recyclable. Provides magic games with natural light with reflections and nuances.
Weaknesses: In hot sunny weathers could be a problem to keep the interior fresh. Fragile. It is a cold material.

A: 장점-유리와 같은 재료는 없으며 내부-외부 상황을 해결하는데 대부분 건물에 필수불가결하다. 100% 재활용이 가능하다. 반사와 뉘앙스가 있는 자연광으로 마법 같은 효과를 부린다.
단점-더운 날씨에는 실내를 쾌적하게 유지하는 것이 어려울 수 있다. 부서지기 쉽고 차가운 물질이다.

WOOD-Q1: Tell us about your favourite project that you used wood in or another architect's work - interior, facade, etc.

A: I did not work with wood as main material for a building. In our dry and too sunny weather in Spain, wood is not a material very common in architecture, the sun can destroy it if you use it in exteriors. Wood has been used traditionally only in structure for ancient constructions.
I have worked with wood in our refurbishment projects where the slabs or roof structures are made with this material, so my experience with wood is more related to building rehabilitation

A: 나는 목재를 건물의 주재료로 사용한 적이 없다. 스페인의 건조하고 매우 밝은 날씨에서 나무는 매우 흔한 건설 재료가 아니며, 외부에 사용하면 태양에 손상된다. 목재는 전통적으로 고대 건축물의 구조에서만 사용되었다.
나는 목재로 된 슬래브나 지붕 구조물을 보수하는데 이 재료를 쓴 적이 있다. 그래서 목재에 대한 나의 경험은 건물의 보수와 물리적 특성을 복구하는 목재의 처리 방법에 연관된다.

and how to treat the wood to recover its mechanical properties.

WOOD-Q2: What are the strengths and weaknesses of wood?

A: Strengths: great structural behavior, is a recyclable material and its production can be sustainable if you work with wood from sustainable forests (FSC forests). Is the only material that produces oxygen when is being produced (growing).
Weaknesses: Can be affected by parasites, fire, and is not the best material for dry and hot weathers like the one I use to work in.

A: 장점-훌륭한 구조적 속성이 있고 재활용 가능한 재료이며 지속 가능한 숲 (FSC 숲)의 목재로 작업하면 그 생산 역시 지속 가능하다. 생산(자라는) 과정에서 산소를 생산하는 유일한 재료이다.
단점-기생충의 영향을 받을 수 있으며, 불에 약하고, 내가 일하는 건조하고 더운 날씨에 쓰기에 좋은 재료는 아니다.

Piuarch

Who is ...?

Francesco Fresa, Germán Fuenmayor, Gino Garbellini and Monica Tricario formed the Piuarch studio in 1996 out of a desire to merge different experiences into a shared architectural project.

The studio is located in an open space in a former industrial building that once hosted a typography business in Brera, in the centre of Milan.

Here, Piuarch designs public buildings, office and residential complexes, commercial spaces, boutiques, shopping malls and even urban plans, with the contribution of consultants from various disciplines.

Q1: What is material to an architect
      (or to you)?

A: Architecture is made of material, and like every material it's susceptible to the changes of time. This is the reason that leads us to design guaranteeing not only an aesthetic balance, but also the durability and performance of the materials we use. **We achieve this through a careful research phase to understand how to use them and adapt them according to the aims.** For us, this means innovating through the reinterpretation of materials, with particular attention to natural ones, the properties and qualities of which we seek to enhance.

A: 건축은 재료로 만들어지며 모든 재료와 마찬가지로 시간의 흐름에 취약하다. 이것이 우리가 심미적 균형뿐만 아니라 사용하는 재료의 내구성과 성능을 보장하는 디자인을 하려는 이유이다. **우리는 신중한 연구 단계를 거쳐 재료를 어떻게 사용할지 이해하고 목표에 따라 적용하여 이를 달성한다.** 이는 우리가 향상하려는 자연적인 재료의 특성과 속성에 특별한 주의를 기울이면서 재료의 재해석을 통한 혁신을 뜻한다.

Q2: Tell us about your favourite
      (or most often used) material and why.

A: We don't have a privileged material, but we make a choice that is related to the context and, at the same time, allows us to experiment, overcoming certain limits, trying to take a step forward each time. Since our beginnings, in 1996, we've tried to introduce into our works materials that are unusual for conventional Italian architecture, starting with the first project

A: 우리에게 특별한 재료는 없지만, 콘텍스트와 관련된 선택을 하는 동시에 특정 한계를 극복하고 매번 한 걸음 앞으로 나아갈 수 있게 실험할 수 있는 재료를 고른다. 1996년 우리 사무소 초반부터 우리는 세스토 산 지오바니(Sesto San Giovanni)의 폴라 하우징(Fola Housing) 프로젝트에서 시작하여 전통적인 이탈리아 건축

an aesthetic balance
심미적 균형

D&G Headquarters ©Alberto Piovano

for Fola Housing in Sesto San Giovanni, where we covered the facades with wood, taking inspiration from the young European experiences of the time, first and foremost the Spanish one. Therefore, **the choice of using materials responds to a need for dialogue with the context but also with the cultural history of the space in which we operate.**

에는 특이한 재료를 소개하려고 노력했다. 당시 유럽의 젊은 건축가들에게서 영감을 얻어 목재로 파사드를 덮었다. 따라서 **재료의 선택은 콘텍스트와의 대화뿐만 아니라 우리가 활동하는 공간의 문화사에 대한 필요성에 답한다.**

Q3: When do you decide the material during the design process and what is your criteria? (e.g. budget, client's preference, design concept, climate, etc.)

A: The choice to prefer one material rather than another one is dictated, from the early phases of the project, by the relationship with the context, both in terms of material and in terms of color. Surely the budget affects but what we do is use materials that are not traditionally luxurious but characterized by extreme attention to detail. Another choice depends on the willingness to employ sustainable solutions, often easily recyclable.

A: 프로젝트의 초기부터 재료 및 색상 측면에서 모두 컨텍스트와의 관계에 의해 다른 재료보다 하나의 재료를 선호하게 된다. 물론 예산이 영향을 미친다. 하지만 우리가 하는 일은 통상적으로 고급스럽지 않지만 디테일에 대한 극단적인 주의로 특색을 나타내는 재료를 쓰는 것이다. 또 다른 선택 기준은 지속 가능한 방안을 사용하려는 의지에 달려 있으며 종종 쉽게 재활용 할 수 있다.

Q4: What are some architectural projects that inspired you regarding brick, tile, wood and/or glass? And why?

A: Every day we browse through numerous industry magazines and surf the major online platforms to see the latest architectural projects around the world. We don't want to mention single buildings, but we can certainly say that we're continually updating ourselves on international experiments.

A: 매일 우리는 수많은 산업 잡지를 뒤지고 주요 온라인 플랫폼을 서핑하여 전 세계의 최신 건축 프로젝트를 본다. 우리는 단 하나의 건물만 언급하고 싶지 않지만, 국제적인 실험을 통해 지속적으로 우리 자신을 업데이트한다고 말할 수 있다.

Q5: Tell us about the materials you are interested in or want to use in your projects right now.

A: As we said, we don't have a material that we use preferentially. The choice of a solution always responds to needs related to the context in which we operate, meaning by "context" the set of aesthetic, environmental and cultural factors of a place or a city.

A: 앞서 말했듯이 우선적으로 사용하는 재료는 없다. 해결책은 항상 우리가 활동하는 맥락과 관련된 요구에 응한다. 즉, "콘텍스트"란 장소나 도시의 미적, 환경적 및 문화적 요인의 집합을 의미한다.

BRICK-Q1: Tell us about your favorite project that you used brick in or another architect's work - interior, facade, etc.

A: We've not worked with exposed bricks very often, but the redevelopment project of the Caproni factory in Milan has allowed us to deal with this material and to enhance its characteristics and aesthetic properties. The brick represents, in this case, the value of industrial archaeology, tells the story of the place, its link with Milan, which has always been a city built with bricks since ancient times. The industrial sheds, covered with metal shed roofs, have brick facades that have been maintained and preserved: they are the memory of the past, in sharp contrast with the new, represented instead by the tower placed between the buildings of the twentieth century and covered with a glass facade and a system of sunshades in anthracite-coloured metal. We worked again with the brick in the project for the competition for the enlargement of the hospital of Tambacounda, in Senegal: there we hypothesized to realize the facades in blocks of raw clay, that the same population could have produced exploiting the local resources and the climate.

A: 우리는 노출된 벽돌을 자주 쓰지는 않지만, 밀라노에 있는 카프로니(Caproni) 공장의 재개발 프로젝트로 이 재료를 다뤄보고 그 특징과 미적 특성을 향상할 수 있었다. 이 경우, 벽돌은 산업 고고학의 가치가 있으며, 고대부터 벽돌로 지어진 도시인 밀라노와의 연결 고리와 그 장소에 대한 이야기를 전한다. 금속 지붕으로 덮인 산업 창고에는 잘 유지되고 보존된 벽돌 외관이 있다. 20세기의 건물 사이에 서서 유리 외관과 무연탄색 금속으로 된 차양 시스템으로 덮인 타워가 상징하는 새로움과는 대조적으로 이 건물은 과거의 기억이다. 우리는 세네갈의 탐바쿤다(Tambacounda) 병원 확장을 위한 공모전에서 벽돌을 다시 연구했다. 여기서 우리는 원시 점토 블록의 파사드를 만들고 같은 지역의 인구가 지역 자원과 기후를 이용해서 이를 생산할 수 있었을 거라고 가정해보았다.

TILE-Q1: Tell us about your favourite project that you used tile in or another architect's work - interior, facade, etc.

A: In our architecture we've never worked with ceramics as the architectural envelope of the project. Ceramics have been used exclusively as a horizontal and vertical coating. On this subject we have often experimented with customised solutions in collaboration with companies. In several occasions, we've designed surfaces to best integrate with the philosophy of the project, restoring a global aesthetic: this is the case, for example, with the concept store for the women's lingerie chain Yamamay, where the ivory square flooring, with geometric inserts in marble ton sur ton, allows you to communicate the ideals of lightness, softness, brightness, geometry and symmetry of the point of sale.

A: 건축에서 우리는 세라믹을 건축의 외피로 사용한 적이 없다. 세라믹은 오로지 수평 및 수직 코팅으로 사용했다. 이 주제를 위해 우리는 종종 여러 회사와 협력하여 맞춤형 방안을 실험해왔다. 여러 차례에 걸쳐 우리는 프로젝트의 철학과 가장 잘 통합할 수 있는 표면을 설계하여 전체적인 미학을 복원했다. 예를 들어, 여성 란제리 체인 야마메이를 위한 컨셉 스토어에서 톤 위에 톤을 써서 상아색 사각형 바닥재와 대리석의 기하학적인 인서트로 이상적인 가벼움, 부드러움, 밝기, 기하학과 매장의 대칭을 전할 수 있었다.

GLASS-Q1: Tell us about your favourite project that you used glass in or another architect's work - interior, facade, etc.

A: Glass with different inclinations, coloured, with specific films for the reflection of the sun, or curved; with

A: 각기 다른 경사가 있는 유리, 색이 있는 유리, 태양을 반사하는 특정한 필름이 있는 유리, 매우 큰 패널 또는

very large panels or in thin sheets. Our projects are a real manual for the use of this material, widely adopted to allow the constant quality of natural lighting within the environments. The Quattro Corti project, in the heart of Saint Petersburg, is certainly worth to mention: the historical facades, which are subject to restrictions, of the two pre-existing buildings conceal four courtyards characterised by four different colours: red, green, grey and blue define the façades of each of the courtyards, made with glass elements placed at different angles that generate a kaleidoscopic effect of great effect. The Ekaterinensky Congress Centre in Krasnodar is also in the executive phase, offering an experimental and innovative use of glass: the front facing the river 'Kuban consists of curved frames that, together with the special treatment of glass, reflects the ripples of water.

얇은 시트 등 우리 프로젝트는 이 재료를 위한 실제 매뉴얼이다. 어떤 환경에서든 일정한 자연광을 위해 널리 사용했다. 상트 페테르부르크의 중심부에 있는 콰트로 코르티(Quattro Corti) 프로젝트는 확실히 언급할 가치가 있다. 기존 두 건물의 전통적인 파사드는 각기 다른 네 가지 색상이 특징인 네 개의 안뜰을 은폐한다. 빨간색, 녹색, 회색 및 파란색은 각기 다른 각도에 위치한 유리 요소로 만들어진 각 안뜰의 외관이 있고 여기서 유리는 굉장한 만화경 같은 효과를 낸다. 러시아 크라스노다르(Krasnodar)의 예카테리넨스키 의회 센터도 실험적이고 혁신적인 유리 사용법을 제공하며 현재 건설 단계에 있다. 쿠반 강을 마주보고 있는 파사드는 특별한 처리가 된 유리와 함께 물결을 반영하는 곡선 프레임으로 구성된다.

WOOD-Q1: Tell us about your favourite project that you used wood in or another architect's work - interior, facade, etc.

A: For the redevelopment project of the old building Latteria Sociale Valtellina,

A: 라테리아 소시알레 발텔리나(Latteria Sociale Valtellina)라

we decided to use a material widely available in mountain areas: wood. We created one single roof overhanging at the front and side, and we used the wood in the opaque cladding of the façade, in the structure, and as the interior covering of the roof.

는 오래된 건물의 재개발 프로젝트를 위해 우리는 산악 지역에서 널리 사용되는 재료인 목재를 사용하기로 결정했다. 우리는 앞면과 옆면에 돌출된 하나의 지붕을 만들었고, 파사드의 불투명한 클래딩, 구조 및 지붕의 내부에 나무를 사용했다.

"EACH MATERIAL MAKES SENSE WHEN APPLIED IN A GIVEN CONTEXT AND EACH CONTEXT PUSHES TO THE SEARCH AND IDENTIFICATION OF PARTICULAR SOLUTIONS."

SLOT STUDIO

Who is ...?

SLOT is an active and interdisciplinary architectural design studio. People from diverse professional disciplines contribute to this project.

Our work has reached a profound understanding of the human needs, enabling us to intertwine the constructive and philosophical sides of building.

As architects, we push design to its ultimate material consequences and aim for cultural connectivity: sense of usability, mathematics of space and a wide aesthetic research.

Q1: What is material to an architect (or to you)?

A: **Material is a powerful instrument in architecture. It is responsible for the sensoriality of a space and defines the character of a building.** Material can be imperceptible but it is never invisible so as architects we are really cautious with this element, it has to be tied to its context and the story we try to tell in our projects. We also see material as an opportunity where we stretch its possibilities and conceive new approaches to architecture. We are now exploiting the potential of traditional materials like concrete, glass or wood to new grounds.

A: 재료는 건축에서 강력한 수단이다. 공간의 감각성을 담당하고 건물의 특징을 정의한다. 특별히 감지하지 못 할수도 있지만 항상 눈에 보이기 때문에 건축가로서 우리는 재료에 굉장히 신중하다. 재료는 콘텍스트와 프로젝트에서 우리가 하고자 하는 이야기에 관련 돼야 한다. 우리는 또 재료를 재료 자체의 가능성을 시험하고 건축을 향한 새로운 접근을 상상할 기회로 본다. 지금은 콘크리트, 유리나 나무 같은 전통적인 재료의 잠재적인 가능성을 개척하고 있는 중이다.

Q2: Tell us about your favourite (or most often used) material and why.

A: We don't have one. We can come up with almost anything as a material for our proposals. Different purposes lead us to different material dimensions. We have used concrete, glass, brick, plastic, fibers, rope, wood and even copper.

A: 딱히 없다. 우리는 기획안을 쓸 때 거의 어떤 것이든 재료로 고려할 수 있다. 각양각색의 목표가 각기 다른 재료로 이끌어 준다. 우리는 콘크리트, 유리, 벽돌, 플라스틱, 섬유, 끈, 나무, 구리까지 써봤다.

Q3: When do you decide the material during the design process and what is your criteria? (e.g. budget, client's preference, design concept, climate, etc.)

A: Normally we choose materials once design principles are defined. Sensitivity and connectivity are our two main concerns. We understand space as the jointure of shape and feelings (objectivity and subjectivity) therefore everything we do regarding materiality derives from there. Although in many cases it is actually our client's desire that pushes this decision we always come up with a twist, with a revelation, for example we have used black concrete in the middle of a gray street, or brick in the middle of a glassy context.

A: 우리는 주로 디자인 원칙이 정의된 다음 재료를 고르며, 주된 관심사는 세심함과 연결성이다. 우리는 공간을 모양과 느낌 (객관성과 주관성)의 접합으로 여기기 때문에 물질성에 관련된 우리가 하는 모든 것은 거기서 시작한다. 많은 경우 의뢰인의 바람이 실제로 재료를 결정하지만 우리는 항상 발상의 전환이나 뜻밖의 발견을 함께 디자인한다. 예를 들면 회색 길 가운데에서 까만 콘크리트를 썼었고 유리가 많은 환경에 벽돌을 쓴 적이 있다.

Sensitivity and connectivity ©Poesia

Tate Modern Switch House ©Jim Linwood

Q4: What are some architectural projects that inspired you regarding brick, tile, wood and/or glass? And why?

A: We tend to understand the "traditional" materials in new uses and configurations, one project that inspires

A: 우리는 새로운 용도와 구성에서 "전통적인"재료를 이해하는 경향이 있다. 우리

SLOT STUDIO 315

us is the Tate Modern Switch House by Herzog & de Meuron in which the brickwork was reinterpreted in the folded latticed façade. MVRDV Crystal Houses in Amsterdam is an extraordinary example of this principle we are talking about, they explored the basis of a glass brick and through experimentation in the structural and fabrication techniques they come up with a glass façade that dissolves into clay. We admire the craftsmanship behind this examples.

가 영감을 받는 프로젝트 중 하나는 헤르조그 앤 드 뫼롱 (Herzog and de Meuron)의 테이트 모던 신관(Tate Modern Switch House)이다. 이 프로젝트에서 벽돌 작업은 접힌 격자 모양의 외관으로 재해석 되었다. 암스테르담에 있는 MVRDV의 크리스탈 하우스 (Crystal House)는 우리가 말하는 이 원리의 뛰어난 예이다. MVRDV는 유리 벽돌의 기초를 탐구하고 구조 및 제조 기술의 실험을 통해 점토로 융해되는 유리 외관을 만들었다. 우리는 이 프로젝트의 장인정신을 존경한다.

Q5: Tell us about the materials you are interested in or want to use in your projects right now.

A: We don't have any particular interest on a material, we wait for the design of the building to speak for itself and reveal what material is best according to the concept of the project. We also keep in mind the sensations we want to display and the expression of the building on its context. **We are always exploring and willing to innovate with the proposals especially when the material becomes the main element of the project.**

A: 우리는 한 재료에 특별한 관심은 없다. 건물의 디자인이 스스로 말하고 프로젝트의 개념에 따라 가장 좋은 재료가 무엇인지 밝힐 때까지 기다린다. 또한 우리가 보여주고 싶은 감각과 콘텍스트에 어울리는 건물의 표현을 염두에 둔다. **우리는 항상 탐구 중이며 특히 재료가 프로젝트의 주요 요소가 될 때 기획 단계의 프로젝트를 통해 혁신하고자 한다.**

BRICK-Q1: Tell us about your favorite project that you used brick in or another architect's work - interior, facade, etc.

A: It is definitely Brick Townhouses. In this case the material became a conceptual accomplice. We reached a special treatment in which both conceptuality and shape merged; Line pursues texture and texture integrates space that is the formula.

A: 단연코 '벽돌 타운하우스'를 꼽겠다. 이 경우 재료가 디자인 컨셉의 공범이 되었다. 디자인 컨셉과 모양이 합쳐지는 특별한 경우였다. 선은 텍스쳐를 추구하고 텍스쳐가 공간을 통합한다. 그것이 공식이다.

Brick Townhouses

BRICK-Q2: What are the strengths and weaknesses of brick?

A: It is a rough hard-sensed material, and this is good for certain purposes but once you are looking to invoke fresh, light, transparency signs it won't be very easy to handle. Presence is definitely on its side, so is traditionality and warmth.

A: 벽돌은 거칠고 단단한 재료이며 특정 목적에는 적합하지만 신선하고 가볍고 투명한 느낌을 원한다면 다루기 쉽지않다. 존재감은 분명히 벽돌의 장점 중 하나이며 전통성과 따뜻함도 있다.

TILE-Q1: Tell us about your favourite project that you used tile in or another architect's work - interior, facade, etc.

A: We designed and built an assembly plant for Quin, the leading automotive wooden parts producer and we needed to achieve a very special aesthetic standard: a trendy-elegant-consistent and refreshing effect. Tiles helped us with this task by covering up the main building's exterior.

A: 우리는 자동차 목재 부품 생산 업체인 퀸(Quin)을 위한 조립 공장을 설계하고 건설했다. 매우 특별한 미적 표준을 달성해야 했는데 트렌디하고 우아하며 상쾌한 느낌을 이루어야 했다. 타일로 본관의 외관을 가려서 이룰 수 있었다.

TILE-Q2: What are the strengths and weaknesses of tile?

A: Tiles are versatile and economic. They can go from simulating to conceptualizing, but one has to be very

A: 타일은 다재다능하고 경제적이다. 시뮬레이션에서 개념화까지 폭넓게 쓸 수 있지만, 허세나 저렴함에 빠질 수도 있

careful when it comes to choosing and justifying the use of it cause you may fall into pretentiousness or cheapness.

기 때문에 타일을 선택하고 사용을 정당화하는 데 매우 조심해야 한다.

GLASS-Q1: Tell us about your favourite project that you used glass in or another architect's work - interior, facade, etc.

A: In our project Pisal even though glass was the least used material it became the main character of it all. A massive concrete structure works around a glass viewer specially designed to overlook an artistic floor. Like a habitable kaleidoscope.

A: 유리가 가장 적게 사용된 재료 임에도 불구하고 유리는 우리 프로젝트 피살(Pisal)의 주인공이다. 거대한 콘크리트 구조물이 예술적인 바닥을 내려다 보도록 특별히 설계된 유리 전망대를 감싸게 되어있다. 마치 거주 할 수 있는 만화경처럼.

GLASS-Q2: What are the strengths and weaknesses of glass?

A: It is very common, and yet it keeps surprising us. Perhaps for its powerful visibility or our constant aim for transparency. There is something religious about glass, it is always associated with light and cleanliness. It appears to be the closest thing to nature (although it is not). Glass will always be appreciated if not needed because it closer to shelter

A: 유리는 매우 흔한 재료지만, 우리를 계속 놀라게 한다. 아마도 유리의 강력한 가시성이나 투명성을 향한 우리의 지속적인 목표 때문일지도 모른다. 유리에는 종교적인 무언가가 있다. 항상 빛과 청결함을 연상하게 한다. 자연에 가장 가까운 그 무언가로 보인다 (실제로 그렇지는 않지만). 유리는 장식보다 거처에 더 가깝기 때문

than it is to ornament. 에 필요하지 않더라도 항상 환영받을 것이다.

WOOD-Q1: Tell us about your favourite project that you used wood in or another architect's work - interior, facade, etc.

A: One of our most recent projects is TOCOMADERA retail store, which literally means touch the wood. It is an interior design project where we had to find the correct expression and balance of the woods that are in display, with different finishes and varnishes on the floor, walls and furniture. It is a project that exudes warmth and that sets wood as an essential element in the handcrafted sense that tries to evoke the products sold

A: 가장 최근의 프로젝트 중 하나는 토코마데라 소매점인데, 문자 그대로 나무를 만진다는 뜻이다. 바닥, 벽 및 가구에 각각 다른 마감과 니스를 사용하여 전시된 나무 상품과 어울리도록 알맞는 표현과 균형을 찾아야했던 인테리어 디자인 프로젝트였다. 따뜻함을 풍기고 상점에서 판매하는 제품을 떠올리게 수공예 느낌의 나무를 필수 요소로 설정한 프로젝트이다.

1. Entrada
2. Barra
3. Expositor
4. Escalera
5. Estantes

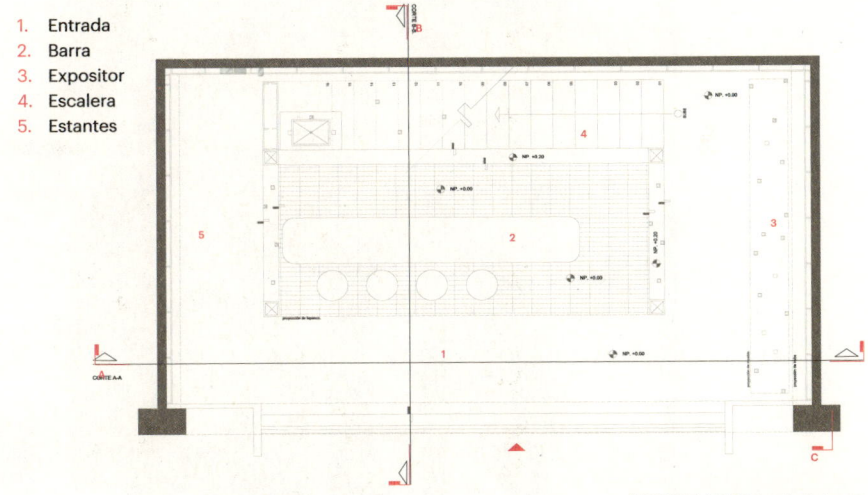

TOCOMADERA retail store

by the store.

WOOD-Q2: What are the strengths and weaknesses of wood?

A: Wood is a very special element that has nobleness, warmth and a sense of craftsmanship. We like to use wood in our projects because it has character and does not require any processes to reveal its charm. Wood can transmit a very wide range of sensations just by playing with its dimensions, malleability, textures and its tonalities; furthermore is a sustainable and acoustic material.

A: 나무는 고귀함, 따뜻함, 장인 정신을 지닌 매우 특별한 재료이다. 나무에는 특별한 개성이 있고 그 매력을 드러내는데 어떤 과정도 필요하지 않기 때문에 우리는 프로젝트에 나무를 사용하는 것을 좋아한다. 나무는 치수, 가단성, 질감 및 음조를 달리 하는 것만으로도 매우 다양한 느낌을 준다. 또한 지속 가능하고 음향적인 소재이다.

"MATERIAL IS RESPONSIBLE FOR THE SENSORIALITY OF A SPACE AND DEFINES THE CHARACTER OF A BUILDING."

SMAR Architecture Studio

Who is ...?

SMAR Architecture Studio, founded by UWA Professor Fernando Jerez (PhD MArch ETSAM) in 2009 and co-directed by Belen Perez de Juan is an awarded Western Australia and Madrid based group of architects and urban thinkers operating with architecture, technology and society. Our projects deploy near-future scenarios as critical instruments for instigating debate about urban and social issues through design and emerging technologies, in order to improve social and political interaction in relation to the environment.

Q1: What is material to an architect (or to you)?

A: We prefer to use the word "matter". We use also the word "substance" in relation to an idea, then the "idea" begins to gain physical properties through its significance. Metaphorically speaking, **ideas can be weighty, have gravitas and become "matter".**
In these linguistic formulations, "ideas" can become "things". Likewise, the "physical" is often an idea as much as it is a thing. The substances we create exist because of the imagination that precedes them. As we impose concepts onto the found condition of the world we alchemise base stuff into a specific material condition. "Ideas", through "matter", can be as real as things can be an imaginary state.

A: 우리는 '물질'이라는 단어를 사용하는 것을 선호한다. 아이디어와 관련하여 "본질"이라는 단어도 사용하며, "아이디어"는 그 중요성을 통해 물리적 특성을 얻는다. 은유적으로 말하면 **아이디어는 무게와 중력을 지닐 수 있으며 "물질"이 될 수 있다.**
이러한 언어적 공식에서 "아이디어"는 "사물"이 될 수 있다. 마찬가지로 '물질'은 종종 사물인 만큼 아이디어이기도 하다. 우리가 창조하는 본질은 그 전의 상상력 때문에 존재한다. 이미 현존하는 상태에 개념을 부과할 때 우리는 기본적인 물질을 특정한 물질로 변화시킨다. "아이디어"는 "물질"을 통해 상상 속 모습 그대로 현실이 될 수도 있다.

Q2: Tell us about your favourite (or most often used) material and why.

A: We are using "Glass" as an important material in our last projects. We are very interested in Transparency, translucency and reflections. Glass is wrapping the spaces forming continuous elevations,

A: 최근 프로젝트에서 '유리'를 중요한 재료로 사용하고 있다. 우리는 투명성, 반투명성, 반사성에 매우 관심이 많다. 여기서 유리는 모서리에서 중단되지 않고 연속적인 입면을

uninterrupted by corners. The visitor flows with the form through a series of interconnected bubbles in the interior and creating interesting reflections with the exterior.

형성하며 공간을 감싼다. 방문자는 내부에 있는 일련의 상호 연결된 "거품"을 통해 건물의 형태와 함께 움직이며 외부와 흥미로운 반사를 만든다.

Q3: When do you decide the material during the design process and what is your criteria? (e.g. budget, client's preference, design concept, climate, etc.)

A: At the beginning, in our office SMAR is all about research, which involves the context, geography, climate, social issues, program, budget, etc... Then the "idea" begins to gain physical properties through its "materialization".

A: 처음에 우리 SMAR는 콘텍스트, 지리, 기후, 사회 문제, 프로그램, 예산 등을 포함하는 연구로 프로젝트를 시작한다. "아이디어"는 "물질화"를 통해 차차 물리적 특성을 얻는다.

Q4: What are some architectural projects that inspired you regarding brick, tile, wood and/or glass? And why?

A: Maison d'verre by Pierre Chareau in Paris. Farnsworth House by Mies van der Rohe and of course SANAA's Glass Pavilion in Toledo, Ohio.
Regarding bricks and tiles for the project in Yadz, the work of Guastavino in USA in XIX Century has been a great

A: 파리에 있는 피에르 샤로(Pierre Chareau)의 메종 드 베르(Maison de Verre). 미즈 반 데어 로에의(Mies van der Rohe)의 판스워즈 하우스(Farnsworth House). 그리고 물론 미국 오하이오주 톨레도에 있는 SANAA의 글래스 파빌리온(Glass Pavilion). 야드즈(Yadz) 프로젝트에

Farnsworth House ©Benjamin Lipsman

Science Island and Innovation Center in Lithuania

SMAR Architecture Studio 327

influence.

Q5: Tell us about the materials you are interested in or want to use in your projects right now.

A: We always try to use materials which are available locally and also that fit in the budget. For the "Faculty of Biomedical Sciences" that we are designing in Madrid, because the budget is very tight we are using concrete and polycarbonate. For the "Science Island and Innovation Center in Lithuania", we are using concrete, local stone paving and low iron glass.

TILE-Q1: Tell us about your favourite project that you used tile in or another architect's work - interior, facade, etc.

A: We have used tiles in our project for Yadz, using a traditional technique that allows the vaults to be supported with the minimum structure. The work of Gaudi in Barcelona, and Guastavino in USA in XIX Century have been a great influence.

서 벽돌과 타일을 쓸 때 19세기 미국 건축가 구아스타비노(Guastavino)의 작업에서 큰 영향을 받았다.

A: 우리는 항상 현지에서 구할 수 있는 동시에 예산에 맞는 재료를 사용하려고 노력한다. 마드리드에 설계하고 있는 "생물의학부" 건물의 경우 예산이 매우 빠듯하기 때문에 콘크리트와 폴리카보네이트를 썼다. "리투아니아 과학 섬 및 혁신 센터"의 경우 콘크리트, 지역 석재로 만든 포장재 및 저철분 유리를 사용했다.

A: 우리는 야드즈(Yadz) 프로젝트에서 타일을 사용했으며, 최소한의 구조로 아치형 천장을 받칠 수 있는 전통 기술을 사용했다. 19세기 바르셀로나의 가우디(Gaudi)와 미국의 구아스타비노(Guastavino)의 작품에서 큰 영향을 받았다.

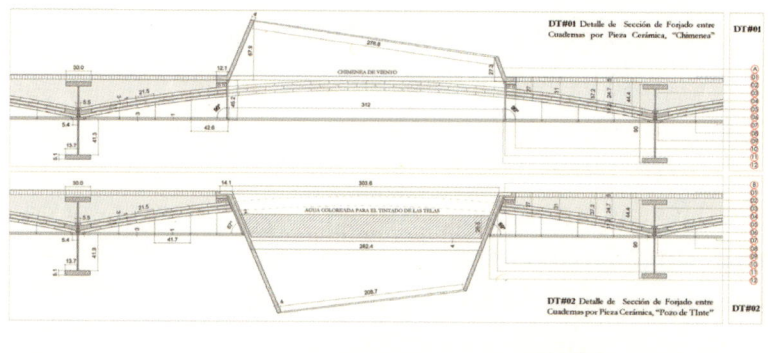

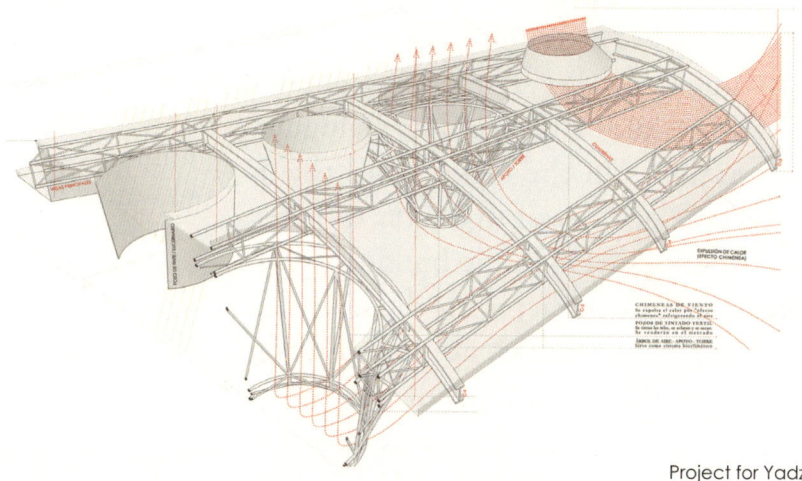

Project for Yadz

TILE-Q2: What are the strengths and weaknesses of tile?

A: Tiles provide infinite compositional possibilities, they are colorful, durable and easy to replace. The weakness is probably related to fragility and the issue that is a traditional technique and they have to be placed one by one by hand.

A: 타일은 무한한 구성 가능성을 지녔고, 색상이 다채로우며, 내구력이 있고 교체하기 쉽다. 단점은 아마도 부서지기 쉬운 점, 전통 기술, 그리고 손으로 하나씩 배치해야 한다는 점이다.

GLASS-Q1: Tell us about your favourite project that you used glass in or another architect's work - interior, facade, etc.

A: SMAR's current project for "Science Island and Innovation Center in Lithuania", currently to be completed in 2020 it is a project that celebrates transparency. The 300m long glass wall, with Low iron super white panels of 8x1.60m creates a façade that dialogues with nature. On the other hand the 3 courtyards of 27m of diameter of glass separates the spaces in the building, give visitors visual contact with the outside, the Museum activities, and exhibition spaces, at all times. One of the most interesting things in this project is the way it creates an intimate relationship between the inside and the outside, giving visitors the feeling of walking under the sky...feeling the green atmosphere of the island.

A: SMAR의 현 프로젝트인 "리투아니아 과학 섬 및 혁신 센터"는 현재 2020년에 완료될 예정이며 투명성을 기념하는 프로젝트이다. 가로1.60m, 세로 8m의 매우 하얀 저철분 유리 패널이 있는 300m 길이의 유리 벽은 자연과 대화하는 파사드를 만든다. 반면에 지름 27m의 안뜰 세 곳은 건물의 공간을 분리하고 방문자에게 외부, 박물관 내 활동 및 전시 공간과 시각적 접촉을 항시 제공한다. 이 프로젝트에서 가장 흥미로운 점 중 하나는 내부와 외부 사이에 친밀한 관계를 형성하여 방문자에게 하늘 아래에서 걷는 느낌을 주는 동시에 섬의 녹지 분위기를 느낄 수 있게 하는 것이다.

GLASS-Q2: What are the strengths and weaknesses of glass?

A: Glass weakness is always related to the cost. I'd say that the Strengths are related

A: 유리 단점은 항상 비용과 연관된다. 장점은 물질적 특

to the material properties. Of course, we have to talk here about transparency, reflection and light but glass is also able to be structural. The new Steve Jobs theatre is a recent example of how glass can become the structural element with no frames or pillars. New technologies allow glass also to have insulating properties using very thin layers of low iron glass and argon. We are now developing glass panels of 8m of height for the Science Island Museum we are designing in Kaunas. Lithuania has a very cold climate and we won't have more than 4.5 cm of glass façade.

성과 관련이 있다고 말하고 싶다. 물론 유리의 장점을 논할 때 투명성, 반사, 빛에 관해 이야기해야 하지만 유리는 구조가 될 수도 있다. 새로운 스티브 잡스 극장은 유리가 어떻게 프레임이나 기둥 없이 구조가 될 수 있는지 보여주는 최근 사례이다. 새로운 기술 덕분에 아주 얇은 저철분 유리와 아르곤 층을 사용하여 유리에 단열성이 생겼다. 우리는 현재 카우나스에서 디자인하고 있는 과학 섬 박물관을 위해 높이 8m의 유리 패널을 개발 중이다. 리투아니아는 기후가 매우 추우며 이 프로젝트에서 두께 4.5cm 이상의 유리 파사드는 쓰지 않을 것이다.

Steve Jobs Theater ©Justin Ormont

Stefano Corbo Studio

Who is ...?

Stefano has contributed to several international journals and has published two books: "From Formalism to Weak Form. The Architecture and Philosophy of Peter Eisenman." (Ashgate-Routledge, 2014), and "Interior Landscapes. A visual atlas."(Images, 2016).

In 2012, after working at Mecanoo Architecten, Stefano founded his own office SC-STUDIO (www.scstudio.eu), a multidisciplinary network practicing architecture and design, preoccupied with intellectual, economic and cultural contexts.

Q1: What is material to an architect (or to you)?

A: Materials are one of the many possible tools used by architects in order to express their ideas but, at the same time, materials also offer the possibility to connect people and environment, users and architectures, by provoking psychological and emotional reactions.

A: 재료는 건축가가 아이디어를 표현하기 위해 사용할 수 있는 많은 도구 중 하나이지만 동시에 심리적, 정서적 반응을 자극하여 사람과 환경, 사용자와 건축을 연결할 가능성을 만든다.

Q2: Tell us about your favourite (or most often used) material and why.

A: There is no favourite material in my work. The use of materials depends on the design strategy I have in mind, and of the goals my project aims to achieve. Materials never dictate the formal expression of the project; on the contrary, they are always the result of a holistic approach to design.

A: 가장 좋아하는 재료는 없다. 재료는 내가 염두에 둔 디자인 전략과 내 프로젝트가 달성하고자 하는 목표에 달려 있고 프로젝트의 형태를 좌우하지 않는다. 그 반대로, 재료는 항상 디자인에 대한 전체론적 접근의 결과이다.

Q3: When do you decide the material during the design process and what is your criteria? (e.g. budget, client's preference, design concept, climate, etc.)

A: Materials are decided in a third

A: 재료는 도시 전략의 개념과

moment of the design process, after the definition of an urban strategy, and after its architectural translation. Climate, energy performance and atmosphere normally influence the choice of a certain material, rather than another one.

건축적 번역 후 디자인 과정의 세 번째 순간에 결정한다. 일반적으로 특정 재료를 선택할 때 기후, 에너지 성능 및 대기가 영향을 미친다.

Q4: What are some architectural projects that inspired you regarding brick, tile, wood and/or glass? And why?

A: Apart from the examples mentioned above, **I'm generally inspired by those projects that attempt to challenge the traditional use of materials and explore their hidden potential.** The Swiss firm Herzog & de Meuron started in the 1990s a personal investigation around the skin of the buildings, based on the reinterpretation of conventional materials: famous are those serigraphic patterns that were applied to glass or concrete panels and then moved to the facades of their projects.

A: 앞서 언급한 사례와는 별도로, 나는 일반적으로 전통적인 재료 사용법에 도전하고 숨겨진 잠재력을 탐구하려는 프로젝트에서 영감을 얻는다. 스위스 회사인 헤르조그 앤 드 뫼롱 (Herzog & de Meuron)은 1990년대에 기존 재료의 재해석을 기반으로 건물 외피에 대한 개인적인 조사를 시작했다. 유명한 것은 유리 또는 콘크리트 패널에 적용된 세리 그래픽 패턴이며 그 후 다양한 프로젝트의 파사드가 알려졌다.

Q5: Tell us about the materials you are interested in or want to use in your projects right now.

A: In this moment, I'm particularly interested in polycarbonate. Its application, in fact, allows to work at the same time with transparency, translucency, opacity, ephemerality. Polycarbonate panels can be used in facades, but also in interior spaces as thresholds or partitions.

A: 현재 나는 특히 폴리카보네이트에 관심이 있다. 사실, 폴리카보네이트로 투명성, 반투명성, 불투명도, 일시성을 동시에 작업할 수 있다. 폴리카보네이트 패널은 파사드에 쓸 수 있지만, 내부 공간의 경계 또는 파티션으로도 사용할 수 있다.

Apotheke des Universitätsspitals, Basel(by Herzog & de Meuron)
© Mattes

BRICK-Q1: Tell us about your favorite project that you used brick in or another architect's work - interior, facade, etc.

A: One of the best examples of how brick can be applied to architecture in an evocative and poetic fashion is Sigurd Lewerentz's St. Mark Church in Stockholm, Sweden. This church was built in 1960, and everything is made out of brick: floor, walls, ceiling. Brick and light shape the atmosphere of this sacred space.

A: 벽돌을 연상적이고 시적인 방식으로 건축에 적용할 수 있는 가장 좋은 사례 중 하나는 스웨덴 스톡홀름에 있는 시구르드 레베렌츠(Sigurd Lewerentz)의 세인트 마크 교회(St. Mark Church)이다. 이 교회는 1960년에 지어졌으며 바닥, 벽, 천장을 비롯한 모든 것이 벽돌로 만들어졌다. 벽돌과 빛이 이 신성한 공간의 분위기를 형성한다.

st. Mark Church ©Håkan Svensson Xauxa

BRICK-Q2: What are the strengths and weaknesses of brick?

A: Defining pros and cons in the use of brick is difficult, as every material should be considered in relation to the overall logic of the project, and based on the relationship between its different components.
Bricks are normally associated to a traditional vision of architecture, which is somehow disconnected by technological advances.

A: 벽돌에 대한 장단점을 정의하는 것은 어렵다. 모든 재료는 프로젝트의 전반적인 논리는 물론 다른 요소 간의 관계에 따라 고려되어야 하기 때문이다.
벽돌은 일반적으로 건축의 전통적인 비전이 연상된다. 이 비전은 기술적 진보로 인해 어쩐지 단절되어 버렸다.

TILE-Q1: Tell us about your favourite project that you used tile in or another architect's work - interior, facade, etc.

A: Castelvecchio Museum, a project of adaptive reuse designed by Carlo Scarpa in Verona, Italy, between 1958 and 1974, is a brilliant example of how tiles, along with other materials, can contribute to create a mosaic of colors and spatial effects within the same building.
Tiles, in fact, are always combined with other materials —stone, concrete, wood— and respond to the architect's idea to take the visitor along an intense journey along

A: 1958년부터 1974년까지 카를로 스카파(Carlo Scarpa)가 디자인한 건물 재사용 프로젝트인 이탈리아 베로나의 카스텔 베키오(Castel Vecchio) 박물관은 타일이 다른 재료와 함께 같은 건물 내에서 색상과 공간 효과의 모자이크를 만드는 데 어떻게 기여할 수 있는지에 대한 훌륭한 예이다. 여기서 타일은 사실 석재, 콘크리트, 목재 같은 다른 재료와 항상 결합돼있고 방문객

Stefano Corbo Studio 337

the medieval castle; tiles help emphasize the relationship with the existing structure by analogy or by contrast.

TILE-Q2: What are the strengths and weaknesses of tile?

A: Defining pros and cons in the use of tiles is difficult, as every material should be considered in relation to the overall logic of the project, and based on the relationship between its different components.

GLASS-Q1: Tell us about your favourite project that you used glass in or another architect's work - interior, facade, etc.

A: The Maison de Verre (House of Glass) is a project designed by Pierre Chareau in Paris, from 1928 to 1932. Glass bricks are applied to the façade of the building as a homogeneous translucent wall, with some specific openings to allow transparency and air circulation.
At night, when artificial illumination is on, the house acts as a lighthouse, and is perceived as a dematerialized ephemeral

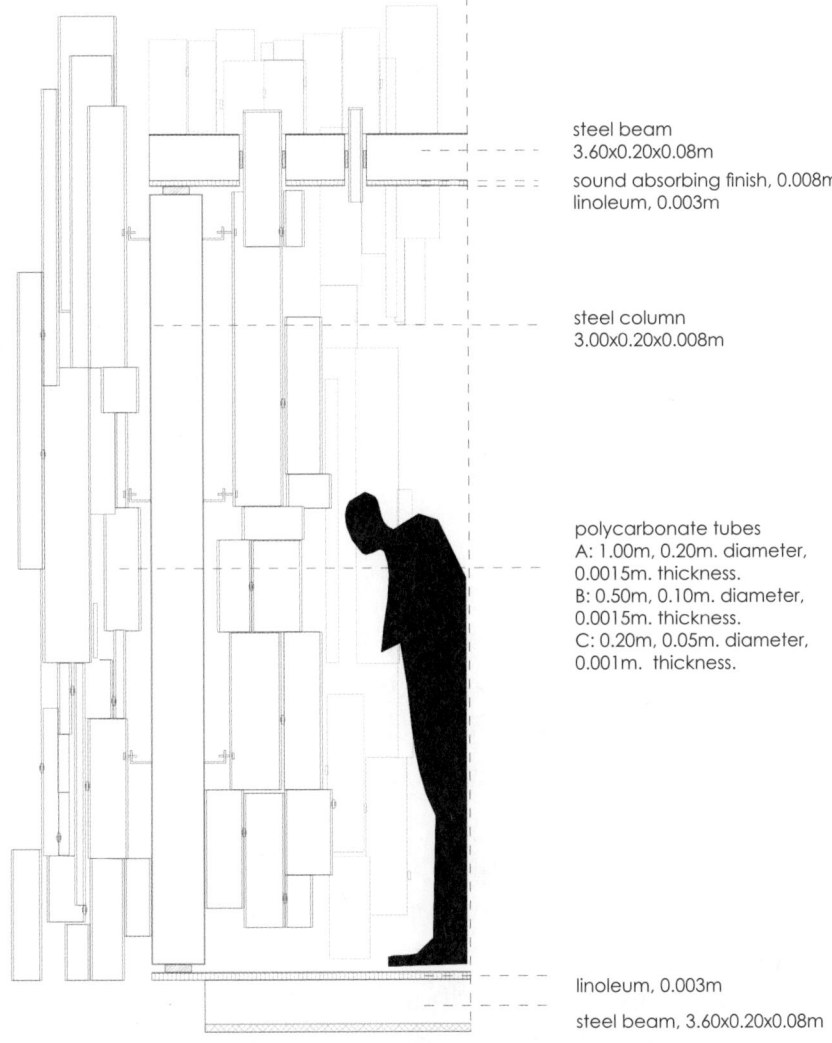

"MATERIALS ARE ALWAYS THE RESULT OF A HOLISTIC APPROACH TO DESIGN."

object.

GLASS-Q2: What are the strengths and weaknesses of glass?

A: Defining pros and cons in the use of glass is difficult, as every material should be considered in relation to the overall logic of the project, and based on the relationship between its different components.

A: 유리에 대한 장단점을 정의하는 것은 어렵다. 모든 재료는 프로젝트의 전반적인 논리는 물론 다른 요소 간의 관계에 따라 고려되어야 하기 때문이다.

WOOD-Q1: Tell us about your favourite project that you used wood in or another architect's work - interior, facade, etc.

A: Sou Fujimoto's Wooden House (2006) is an example of innovative use of wood in architecture. In this project, in fact, there is no separation between floors, walls and ceiling. Its formal genesis derives from the irregular stacking of 350 mm square profile cedar beams.
The end result is a perforated wooden box, which can trigger unusual spatial articulations.

A: 후지모토 소우의 목조 주택(2006년)은 건축에서 나무를 혁신적인 방식으로 사용하는 사례다. 이 프로젝트에서는 사실 바닥, 벽 및 천장 사이에는 분리가 없다. 그 형태는 350mm 정사각형 종단 삼나무 보를 불규칙하게 쌓아서 만들어졌다. 최종 결과는 구멍이 뚫린 나무 상자로 특이한 공간적 표현을 유발한다.

Maison de Verre ©SCSTUDIO

WOOD-Q2: What are the strengths and weaknesses of wood?

A: Defining pros and cons in the use of wood is difficult, as every material should be considered in relation to the overall logic of the project, and based on the relationship between its different components.

A: 나무에 대한 장단점을 정의하는 것은 어렵다. 모든 재료는 프로젝트의 전반적인 논리는 물론 다른 요소 간의 관계에 따라 고려되어야 하기 때문이다.

stpmj

Who is ...?

stpmj is an award winning design practice based in New York and Seoul. The office is founded by Seung Teak Lee and Mi Jung Lim with the agenda, "Provocative Realism". It is a series of synergetic explorations that occur on the boundary between the ideal and the real. It is based on simplicity of form and detail, clarity of structure, excellence in environmental function, use of new materials, and rational management of budget. To these we add ideas generated from curiosity in everyday life as we pursue a methodology for dramatically exploiting the limitations of reality.

Q1: What is material to an architect (or to you)?

A: Material is the layer that defines space and helps to experience with five senses.

A: 공간과 장소를 사람의 오감으로 체험할 수 있도록 도와주는 층위.

Q2: Tell us about your favourite (or most often used) material and why.

A: Brick, we have used brick several times. It maintains its charm over time.

A: 벽돌, 시간이 흘러도 매력을 유지하기때문에.

Q3: When do you decide the material during the design process and what is your criteria? (e.g. budget, client's preference, design concept, climate, etc.)

A: For the exterior, materials are considered as the project started after the site visit. The relation between the site and its surroundings is one of the important starting point, and the overall design direction can be decided based on the composition of materials. For the interior, after schematic design, materials are decided with the detailed consideration of each rooms and their relation.

A: 외부 마감재의 경우 대상지 답사 후, 프로젝트를 시작함과 동시에 고려한다. 대상지 주변과의 관계성이 중요한 시작점이기도 하고 재료의 구성에 따라 디자인의 전체적 방향이 결정되기도 하기 때문이다. 그러나 내부 마감재의 경우는 계획 설계 후 세부적으로 실 구성과 각 실의 관계를 고려하며 결정한다. 건축가가 초기에 재료를 선정해서 진행하는 경우라도 건축주의 예산 상의 제한으로 불가피 하게 바꿔야 하는 경우는 디자인의 방향과 가장 유사한 재료로 중간에 바꾸기도 한다. 건축적 번역 후 디자인 과정의 세 번째 순간에 결정한다. 일반적으로 특정 재료를 선택할 때 기후, 에너지 성능 및 대기가 영향을 미친다.

Q4: What are some architectural projects that inspired you regarding brick, tile, wood and/or glass? And why?

A: Brick_ Colomba Museum/ Peter Zumthor
Delicate use of the brick on the existing building, maximize possibility that brick can create.

A: 벽돌
피터 줌터의 콜럼바 뮤지엄

Q5: What materials were most difficult to handle? And why?

A: Stratum House/stpmj: stratum house is the exposed concrete project that has the various kinds of concrete with different slump/ aggregates and colors. The concrete that has different slump and the ration of aggregates is tricky to control colors. Based on the ratio of water and cement, the amount of dye should be adjusted. It is not easy to predict the color of concrete since the sample from the cement mixer at the site changes its color as it gets hardened. Even with huge difference of slump, the difference of concrete shape was hard to notice with in the thickness of wall. Cleaning the surface after the taking off forms requires some labors.

A: 이천 콘크리트 타설 Stratum House – 각각 다른 슬럼프치와 골재비율을 포함한 콘크리트는 그 조색의 프로세스가 쉽지 않다. 시멘트의 비율과 물의 양에 따라 들어가는 염료의 양이 각각 다르게 책정되어야 하고 레미콘에서 표본으로 현장에서 받은 샘플은 경화되면서 그 색이 다시 변하기 때문에 예측이 어려움. 실제 슬럼프치가 다른 콘크리트가 외벽 두께만큼의 폭을 지닌 거푸집에서 그 변화가 유의미한 정도로 발생하지 않을 때도 있었음. 탈형시 면의 정리해야 하는 양도 상당함.

Kolumba museum by Peter Zumthor

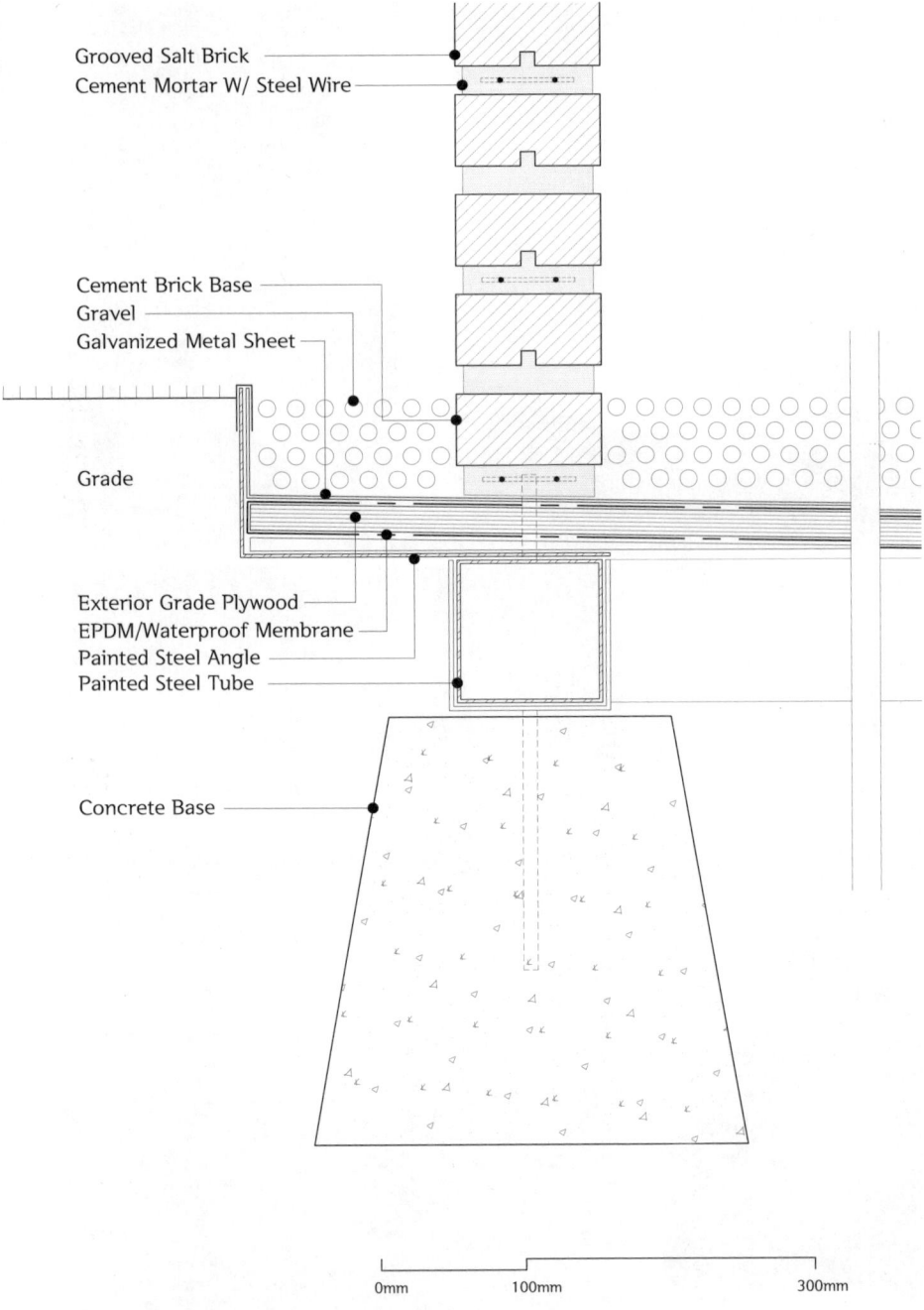

Dissolving Arch Detail

346 Brick, Brick! What do you want to be?

Q6: Have you any memorable episodes about Material?

A: Dissolving Arch, Jeju APMAP exhibition/ stpmj
The pavilion was developed from the old story of Jeju Island. The story is about the structure that had been destroyed by storm every time. It got stable after sacrificing a child to build the structure. We developed the idea of disappearing structure instead of a child by using the weather in Jeju. The salt brick composed arch tunnel and slowly get melted by the rain. And eventually all the brick has disappeared except the grouts that had hold bricks.

A: 제주 APMAP 전시 소금벽돌 제주도의 설화중에 하나를 모티브로 진행한 설치작품이다. 설화의 내용은 제주도의 태풍으로 인해 잘 무너지는 성벽에 아이를 제물로 바치니 그 성벽이 무너지지 않고 잘 버텨주었다는 내용. 우리는 아이의 목숨이 아닌 구조물이 사라지는 내용으로 접근하여, 제주도의 기후를 활용, 소금벽돌로 구조를 만들어 제주도의 비가 잦은 날씨에 소금벽돌이 녹아 사라지게 하는 설치를 하였다.

Q7: Tell us about the materials you are interested in or want to use in your projects right now.

A: Unprocessed wood and stone.

A: 가공 전 나무와 석재.

BRICK-Q1: Tell us about your favorite project that you used brick in or another architect's work - interior, facade, etc.

A: The Masonry / stpmj- by using two different kinds of brick in a single brick façade, it exposed the size and scale differences along with the division of the units.

A: The Masonry / stpmj- 다른 크기의 벽돌을 사용하여 하나의 조적 facade에서 그 스케일 변화가 드러나게 계획되었으며 내부 세대의 구분을 의미하기도 한다.

BRICK-Q2: What are the strengths and weaknesses of brick?

A: Pros: brick assembly can give diverse result of the surface with its various sizes, colors, patterns of assembly, and colors of grouts. Also, it requires relatively low maintenance over time.
Cons: it is not impossible but there is limit to actualize certain size or shape of space with its stack assembly, it is not strong against seismic wave.

A: 장점-사용되는 벽돌의 크기와 색깔, 줄눈의 패턴과 깊이 및 쌓는 방법 등의 세밀한 변화에 따라 전체의 면을 이룰 때 다채로운 입면을 구성할 수 있으며 시간에 따른 재료의 오염도가 안정적이다. 단점-쌓아 올라가는 조적의 방식으로는 (불가능하지는 않지만) 공간의 형태나 크기에 따라 제약이 있고 인장력의 부재로 내진에 취약하다.

TILE-Q1: Tell us about your favourite project that you used tile in or another architect's work - interior, facade, etc.

A: Concert hall and Music school(MUCA)/COR ASOCIADOS
A simple treatment with ceramic tile as exterior finish, provides various outlooks depending on its reflection and where

A: Concert hall and Music school(MUCA)/COR ASOCIADOS
세라믹타일로 마감하여 보는 위치에 따라 건물의 외관이 달리 보이는 효과를 내고 있다.

Under Construction

you stand.

TILE-Q2: What are the strengths and weaknesses of tile?

A: Pro: tile can be applied various cases, it is light and has wide variety of selection. Con: for the use of exterior, some loss or crack could be happened because of the contraction and expansion of the material itself as exposed on sunlight and water directly.

A: 장점-타일은 디자인 적용의 폭이 넓다. 제품의 다양성이 확보되어 있고 가볍다. 단점-외기에 면하는 경우 기후에 따라 제품 자체가 갖는 재료의 수축 팽창에 영향을 주기 때문에 탈락 및 제품의 하자로 이어질 확률이 있다.

GLASS-Q1: Tell us about your favourite project that you used glass in or another architect's work - interior, facade, etc.

A: Glass Farm/MVRDV-brick printed glass construction the original materiality of glass, transparency transformed to a new materiality of glass by printing the images on the glass.

A: Glass Farm/MVRDV 글래스를 개념화하는 작업이 가능했다

GLASS-Q2: What are the strengths and weaknesses of glass?

A: Pro: Transparency for views and the

A: 장점-채광이나 조망을 위

light, opacity for privacy, it can control the opacity with frit or print pattern.
Con: cost / price / Fragility

WOOD-Q1: Tell us about your favourite project that you used wood in or another architect's work - interior, facade, etc.

A: Shear House/stpmj-the project of wood frame structure with wood finish. Sliced and shifted gable roof in one monolithic wood material.

WOOD-Q2: What are the strengths and weaknesses of wood?

A: Pros: natural, light, easy to install. Environmental-friendly-material that stimulate touch, smell, and as well as sight simultaneously.
Cons: The consideration of contraction and expansion of the material and the continuous maintenance are required.

한 투명함, 반대로 프라이버시를 위한 불투명함. 단일 재료임에도 투명함 정도를 여러 테크놀로지를 통해 조절이 가능하다. Frit 이나 프린트 패턴의 다양함 확보
단점-예산에서의 비중이 크다 (코스트), 깨지기 쉬워 유지보수에 비용이 발생한다.

A: Shear House /stpmj- 경량 목구조에 외부 마감까지 목재로 진행했던 프로젝트로 틀어진 박공지붕의 타이폴로지를 단일 재료인 나무로 디자인.

A: 장점-가볍고 시공이 용이하며 자연친화적이다. 어떤 처리를 하느냐에 따라 시각적인 효과와 더불어 촉각적, 후각적 감각들을 동시에 느끼게 할 수 있는 천연재료
단점-기후에 따라 수축 팽창에 따른 변위에 대한 고려가 설계 및 시공 시 요구된다.

Studio Farris Architects

Who is ...?

Studio Farris Architects is an architectural practice based in Antwerp, Belgium, founded by Italian architect Giuseppe Farris in 2008.

The studio's goal is to discover the intrinsic potential in every project, questioning the obvious, exploring the surroundings and cultural heritage.

Architecture, interior architecture, furniture design, lighting and graphic design are all part of the same overall concept and are designed as a coherent whole, with particular attention to detail.

Q1: What is material to an architect
   (or to you)?

A: Material is endless.
There are a thousand different possibilities in one material alone.
Material becomes an expressive and a powerful architectural element only through the creativity of the designer.

A: 재료에는 끝이 없다. 한 가지 재료만 해도 천 가지의 가능성이 있다. **재료는 디자이너의 창의력을 통해서만 표현력이 넘치고 강력한 건축 요소가 된다.**

Q2: Tell us about your favourite
   (or most often used) material and why.

A: I don't have a favourite material. I don't think that there is a good material or a bad material.
Each material has its own physical property and it depends on the architect to treat it with sincerity.

A: 좋아하는 재료는 없다. 좋은 재료나 나쁜 재료가 있다고 생각하지 않는다. 각 재료는 각기 물리적 특성이 있으며 건축가가 진정성을 가지고 다루는 데 달려있다.

Q3: When do you decide the material during the design process and what is your criteria? (e.g. budget, client's preference, design concept, climate, etc.)

A: There is not a specific moment to decide what material we will use for a project. It depends. Sometimes it is during the design concept, sometimes in a later

A: 프로젝트에 어떤 재료를 사용할지 결정하는 구체적인 순간은 없다. 다 다르다. 때로는 디자인 콘셉트 중에, 때로는 후기 단계에 결정한다. 프로젝

phase. It's a process that evolves with the project.

트와 함께 진화하는 과정이다.

Q4: What are some architectural projects that inspired you regarding brick, tile, wood and/or glass? And why?

A: I like Peter Zumthor's sound pavilion in wood. The use of wood is both sensorial and functional.

A: 나는 피터 줌터(Peter Zumthor)의 나무로 된 음향 파빌리온이 마음에 든다. 나무를 감각적이면서도 기능적으로 사용했다.

Q5: Tell us about the materials you are interested in or want to use in your projects right now.

A: I'm working on different projects with different environmental surrounding, context and programs.
Consequently different materials and different uses of the same material.
We will use bricks, metal, glass, wood.

A: 현재 환경, 상황, 프로그램을 달리하는 다양한 프로젝트를 진행하고 있다. 결과적으로 각기 다른 재료와 동일한 재료의 다른 용도이다. 우리는 벽돌, 금속, 유리, 나무를 사용할 것이다.

BRICK-Q1: Tell us about your favorite project that you used brick in or another architect's work - interior, facade, etc.

A: We used bricks for the project of the

A: 우리는 벨기에의 앤트워프

Antwerp Zoo and for the Farmhouse Lennik project.

동물원과 레니크 농장 프로젝트에 벽돌을 사용했다.

BRICK-Q2: What are the strengths and weaknesses of brick?

A: The most interesting advantages are: It is economical, it is hard and durable, it is low maintenance, reusable and recyclable, it produces less environmental pollution during the manufacturing process.
The disadvantages could be: construction with bricks is time consuming. Since bricks absorb water easily, they cause flaking when not exposed to air.

A: 가장 흥미로운 장점은 경제적이며, 단단하고, 내구성이 있으며, 유지 보수가 낮고, 재사용 및 재활용을 할 수 있으며, 제조 과정에서 환경 오염이 적다는 점이다. 단점은 벽돌로 건축하는 것은 시간이 많이 소요되고, 물을 쉽게 흡수하기 때문에 공기에 노출되지 않으면 박리가 생긴다는 점이다.

GLASS-Q1: Tell us about your favourite project that you used glass in or another architect's work - interior, facade, etc.

A: We used glass for the Flemish parliament pavilion for the Park tower in Antwerp, for the City Library in Bruges.

A: 우리는 플랑드르 의회 파빌리온, 앤트워프에 있는 공원 타워, 그리고 브뤼헤에 있는 도시 도서관에 유리를 사용했다.

GLASS-Q2: What are the strengths and weaknesses of glass?

A: Advantages could be: transparency and translucency, light weight, hard and durable, reusable and recyclable. Disadvantages could be: costly material, requires regular cleaning.

A: 장점은 투명성과 반투명성. 경량. 단단함과 내구성. 재사용 및 재활용 할 수 있다는 점이다. 단점은 가격이 비싸며 정기적인 청소가 필요한 점이다.

Antwerp Zoo ©Martino Pietropoli

SUPA architects
schweitzer song

Who is ...?

SUPA is working within the narrow intersection of conceptual design, art, theory and education.

They are focused on the conceptual approach towards architectural design, finding new ways to expand the vocabulary of architecture. Their special interest lies in the exploration of the specific sociocultural context inherent in a task as the interface between everyday life, art and architecture.

Q1: What is material to an architect (or to you)?

A: We would consider ourselves as "material fundamentalists" in the sense that material is not a choice but the result of the architectural process. Material should not be applied, it should not be random or exchangeable, but be the inevitable consequence of our design process. **We try to give material in our buildings a meaning by making them structural and at the same time visible as structure.**

A: 재료를 선택하는 게 아니라 건축 과정의 결과라는 의미에서 우리 자신을 "재료 근본주의자"라고 생각한다. 재료는 그냥 적용되거나 무작위여서는 안되며, 대체할 수 없어야 하고, 설계 과정의 필연적인 결과여야 한다. **우리는 우리 건물의 재료를 구조적으로 쓰는 동시에 그 구조가 눈에 보이게 하여 의미를 부여하려고 노력한다.**

Q2: Tell us about your favourite (or most often used) material and why.

A: Concrete and wood. Both materials can solve everything from structure to finishing, from aesthetics to the quality of life within them.

A: 콘크리트와 나무. 두 가지 재료 모두 구조에서 마무리까지, 미학에서 삶의 질에 이르기까지 모든 것을 해결할 수 있다.

Q3: When do you decide the material during the design process and what is your criteria? (e.g. budget, client's preference, design concept, climate, etc.)

A: It is always a mix of many consi-

A: 항상 많은 고려 사항이 섞여

SUPA architects schweitzer song 359

derations. Of course budget and the client's input is a very dominant factor but all in all we are trying to express a "truth" about the building through material. Using material as decoration or in a decorative fashion is a red line for us.

있다. 예산과 의뢰인의 의견은 물론 매우 중요한 요소이지만, 우리는 대체로 재료를 통해 건물에 대한 "진실"을 표현하려고 노력한다. 장식으로 또는 장식적인 방식으로 재료를 사용하는 것은 우리의 한계점이다.

Q4: What are some architectural projects that inspired you regarding brick, tile, wood and/or glass? And why?

A: We are fascinated by projects where material and structure and function and aesthetics become one, when all these aspects are not separable anymore. Very radically this can be seen in projects like Christian Kerez' House with one Wall, in Junya Ishigami's KAIT Workshop, or even Jo Nagasaka's renovation projects.

A: 우리에게는 재료, 구조, 기능과 미학이 하나가 되는 프로젝트가 매혹적이다. 크리스천 케레즈(Christian Kerez)의 벽 하나의 집, 이시가미 준야(Junya Ishigami)의 KAIT 워크숍, 심지어 나가사카 조(Jo Nagasak)의 레노베이션 같은 프로젝트에서도 확연히 볼 수 있다.

KAIT Workshop © Epiq

Q5: What materials were most difficult to handle? And why?

A: This very much depends on the skill level and experience of the local construction companies. We ran into problems with very common materials and had no problems with highly complicated materials, but it was never the fault of the material itself.

A: 이는 지역 건설 회사의 기술 수준과 경험에 달려있다. 우리는 매우 흔한 재료를 쓸 때 문제가 생긴 적이 있지만, 매우 복잡한 재료 때에는 아무런 문제도 없었던 경우가 있었다. 하지만 절대 재료 자체의 잘못인 적은 없었다.

Q6: Have you any memorable episodes about Material?

A: On every building site we had there is a moment during the construction when everything still is in chaos when suddenly the material starts to awake, when one suddenly becomes aware of it. This awareness very much is in relation to suddenly feeling the space in all its consequences. It is a magical moment when all pieces fall into one.

A: 우리가 손댔던 모든 건물 현장에서 건설 중 한순간 모든 것이 혼란에 빠져있는 와중 갑자기 재료가 깨어나기 시작하고 그 재료가 인식되는 순간이 있다. 이 인식은 갑자기 모든 결과에서 비롯된 공간을 느끼는 것이고, 이는 모든 조각이 하나로 떨어지는 마법의 순간이다.

Q7: Tell us about the materials you are interested in or want to use in your projects right now.

A: Since ages we try to apply plastic

A: 우리는 오랜 시간 동안 건

materials in our buildings; already existing materials just taken out of context and used for architecture. We did experiments with structural EPS blocks, PVC coated tarpaulin fabrics, RPP (reinforced PVC panels) but unfortunately it is very hard to convince clients to use non-established materials.

물에 플라스틱 재료를 적용하려고 노력해왔다. 이미 현존하는 재료를 콘텍스트가 아니라 건축에 사용하는 것이다. 우리는 구조적 EPS 블록, PVC 코팅 타폴린 직물, RPP(강화 PVC 패널)를 실험해봤지만, 불행히도 의뢰인에게 새로운 재료를 사용하도록 설득하는 것은 매우 어려운 일이다.

BRICK-Q1: Tell us about your favorite project that you used brick in or another architect's work - interior, facade, etc.

A: Brick is most fascinating for us when it still was used structurally and visibly as such. Unfortunately brick today became a finishing material, an aesthetic surface without deeper meaning as an architectural element. Basically today the brick is reduced to a tile.

A: 벽돌은 구조적으로, 혹은 구조로서 눈에 띄게 사용되었을 때 가장 매혹적이다. 불행히도 벽돌은 오늘날 건축 요소로서의 깊은 의미가 없는 미적 표면 뿐인 마감재가 되었다. 한마디로 오늘날 벽돌은 타일이 되어버렸다.

BRICK-Q2: What are the strengths and weaknesses of brick?

A: A brick is very raw and fundamental. It is a minimal building block out of which almost everything can be generated.

A: 벽돌은 매우 노골적이고 근본적이다. 거의 모든 것을 생성 할 수 있는 최소한의 재료이다.

TILE-Q1: Tell us about your favourite project that you used tile in or another architect's work - interior, facade, etc.

A: **A tile is the little pretty brother of the brick.** It is still the best solution to certain functional problems within a building (e.g. in the bathroom). We try to apply it very reduced, white, minimalistic.

A: 타일은 벽돌의 예쁜 동생이다. 이것은 여전히 건물 내의 화장실과 같은 특정 기능적 문제에 대한 최선의 해결책이다. 우리는 타일을 흰색에 매우 간결하게 사용하려고 한다.

TILE-Q2: What are the strengths and weaknesses of tile?

A: We are dreaming of a seamless tile, the beautiful surface of a ceramic tile without the joints. If it would be possible to bake and glaze a whole bathroom or kitchen we would be the first to apply it.

A: 우리는 매끄러운 타일, 조인트가 없는 아름다운 세라믹 타일 표면을 꿈꾼다. 만약 화장실이나 부엌 전체를 구워서 유약을 칠 수 있다면, 우리가 제일 먼저 쓸 것이다.

GLASS-Q1: Tell us about your favourite project that you used glass in or another architect's work - interior, facade, etc.

A: Glass we would almost not consider a "material" in architecture but an unavoidable necessity of architecture. Again, for us glass is most fascinating

A: 우리 생각에 건축에서 유리는 거의 '재료'라고는 할 수 없지만, 건축에서 불가피하게 필요하다고 본다. 다시 말하지만, 우리에게 유리는 실험

when it is used structurally like in experimental pavilions or frameless just as itself.

파빌리온처럼 구조적으로 사용되거나 프레임이 없는 경우 가장 매혹적이다.

GLASS-Q2: What are the strengths and weaknesses of glass?

A: Glass doesn't have any weakness, frames have. A frame always feels like a compromise.

A: 유리에 단점은 전혀 없다. 창틀은 언제나 타협처럼 느껴진다.

"HOW WE USE MATERIAL SAYS WHO WE ARE."

House P ©김희천

SUPA architects schweitzer song 365

WOOD-Q1: Tell us about your favourite project that you used wood in or another architect's work - interior, facade, etc.

A: Wood to us is endlessly fascinating. It is the most fundamental material in architecture. Both of us were socialized by wood. Ryul was raised in a Korean Hanok, I was raised in an Upper-Austrian farmhouse. This experience still dominates our thinking, our aesthetics, and our architectural strategies. Although both houses seem to be physically so far apart, philosophically they provided a common ground between our cultures for us to work with.

A: 우리에게 나무는 끝없이 매혹적이며 건축에서 가장 근본적인 재료이다. 우리 둘 다 나무에 둘러싸여 자랐다. 률은 한국의 한옥에서 자랐고, 나는 오스트리아 오버외스터라이히(Oberösterreich) 주의 한 농가에서 자랐다. 이러한 경험은 여전히 우리의 사고, 미학, 그리고 건축적 계획을 지배한다. 이 두 집 모두 물리적으로 멀리 떨어져 있는 것처럼 보이지만 철학적으로 우리 문화 사이에 공통된 기반을 제공했다.

WOOD-Q2: What are the strengths and weaknesses of wood?

A: The overwhelming strength of wood as a building material is its metaphysical quality. Wood lives and reflects our own lives when living within it. It ages, it moves, it makes sounds, it warms, it scents, it changes.

A: 건축 자재로서 나무의 압도적인 장점은 형이상학적 특성이다. 나무 안에서 살 때 나무는 우리 자신의 삶을 반영하며 같이 살아간다. 나이를 먹고, 움직이고, 소리를 내고, 따뜻하고, 냄새가 나고, 변화한다.

TAKK Architecture

Who is ...?

Takk is a space for architectural production focused in the development of experimental and speculative material practices in the intersection between nature and culture in the contemporary framework, with a special attention on the overcoming of anthropocentrism on its different ways(political, ecological, cultural, on gender), and on the definition of new notions of beauty through the articulation of the difference by assembling a multiplicity of materials from different origins and conditions.

Q1: What is material to an architect (or to you)?

A: For us, political, economical, aesthetical or gender issues are constantly discussed around the raw material and the learning of their use protocols.

A: 우리는 원자재와 그것들의 사용 방법에 대해 연구하기 위해 정치적, 경제적, 미적 또는 성별 문제를 끊임없이 논의한다.

Q2: Tell us about your favourite (or most often used) material and why.

A: We don't have a more often used material, but indeed we have a clear idea on how to look for the. **A clear example of material mediation more or less technified are some of the places that we visit frequently for the construction of our architecture.** Spaces such as Do-it-yourself centers, hardware stores, haberdasheries, flower shops, or even stationery shops, for us not only they work as another kind of place committed to the sale of products, but we mainly consider them as meeting and discussion places, in the end, spaces of production of subjectivity.

A: 특별히 더 자주 사용하는 재료는 없지만, 실제로 알맞은 재료를 어떻게 찾아야하는지에 대한 명확한 아이디어가 있다. **우리가 거의 기술화한 재료 선택의 명확한 방법은 우리가 건축을 위해 자주 방문하는 장소에 있다.** DIY 센터, 하드웨어 매장, 잡화점, 꽃가게, 심지어 문구점 같은 공간은 상품 판매에 투입되는 종류의 장소일 뿐 아니라, 주로 회의와 토론 장소로 간주하고 결과적으로 주관성을 생산하는 공간으로 여긴다.

Q3: When do you decide the material during the design process and what is your criteria? (e.g. budget, client's

> We are interested in every material that is well known by "non-experts"

TAKK MOUNTAIN

preference, design concept, climate, etc.)

A: Our approach to architecture, or our references as architects come from spaces that we could consider as perihperal if we think in classical disciplinary terms. We work from the observation of particular cases that we afterwards try to bring to our more general architectural practice. Because of this sometimes our work could be sometimes related to cultural or anthropological studies, and our projects are very different from each other or are developed in very different formats.

Issues as for example, how to prepare a christmas table, or its incorporated aesthetics, the interior design magazines, a macramé workshop, parties, the political role of food, the use of materials or the color in the market economy, are situations from which architecture can learn as all these issues are material environments more or less technified, that can be constantly reevaluated by those societies that take part on them.

A: 건축에 대한 우리의 접근 방식이나 건축가로서의 참고 자료는 우리가 고전적인 측면에서 생각할 때 주변적인 공간으로 간주할 수 있는 곳에서 비롯된다. 우리는 특정 사례를 관찰하고 이후에 좀 더 일반적인 우리의 건축 작품에 도입하려고 시도한다. 이 때문에 우리 프로젝트는 때로 문화적 또는 인류학적 연구와 관련이 있을 수 있으며, 서로 매우 다른 형식으로 개발되거나 다른 형태로 진행된다. 예를 들어, 크리스마스 저녁상을 준비하는 방법이나 그에 따른 미학, 인테리어 디자인 잡지, 마크라메 워크숍, 파티, 음식의 정치적 역할, 재료의 사용 또는 시장 경제에서의 재료나 색채와 같은 문제는 건축이 배울 수 있는 상황이다. 이러한 모든 문제는 거의 기술화된 물질적인 환경이며 이에 참여하는 사회에 의해서 끊임없이 재평가될 수 있다.

Q4: Tell us about the materials you are interested in or want to use in your projects right now.

A: We are interested in every material that is well known by "non-experts", but that is not common to use it as an architectural material.

A: 우리는 일반적으로 사용하지 않는 "비주류" 재료에 관심이 많다. 하지만 이것들은 평소에 건축재료로 사용되지 않는다.

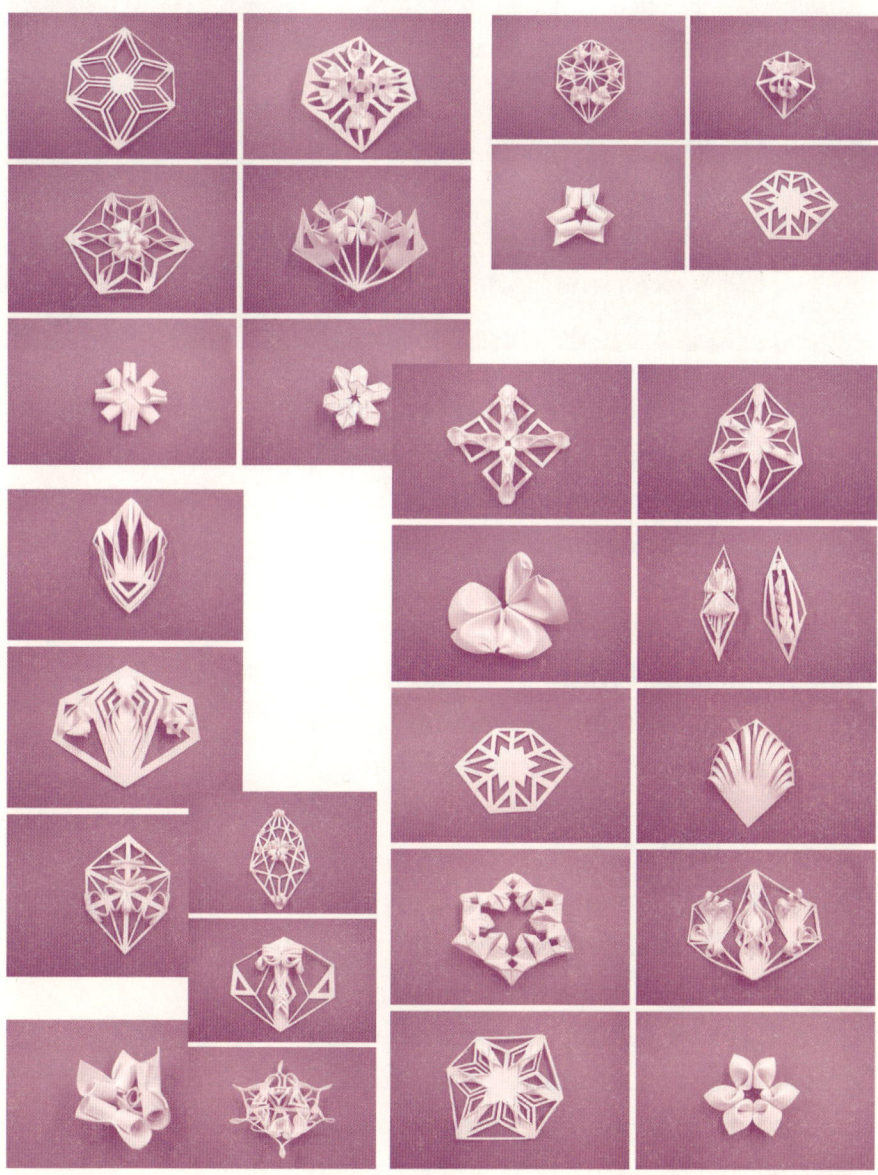

tests for the tiles

Hospital de Sant Pau ©Nicky Boogaard

BRICK-Q1: Tell us about your favorite project that you used brick in or another architect's work - interior, facade, etc.

A: Sant Pau Hospital in Barcelona by Lluís Domenech I Montaner is an icon of the modernist catalan movement which inspires us everyday living in Barcelona.

A: 루이스 도메네 이 몬타네(Lluís Domenech i Montaner)가 지은 바르셀로나의 산트 파우 병원(Sant Pau Hospital)은 바르셀로나에서 매일 생활하는 우리에게 영감을 주는 카탈로니아 모더니스트 운동의 아이콘이다.

BRICK-Q2: What are the strengths and weaknesses of brick?

A: Brick is a very cheap material, that can be used both as a structure or as a façade, that is something we find very interesting. On the other hand it is very linked

A: 벽돌은 매우 값싼 재료로, 구조나 외관으로 사용할 수 있으며 이는 우리가 매우 흥미롭게 생각하는 점이다. 반면에 오늘날 벽돌은 우리 주

nowadays in our context to a certain architectural style with certain political implications that are not exactly followed by us.

변에서 우리가 특별히 따르지 않는 특정 정치적 의미를 지닌 특정 건축 양식과 매우 연관되어있다.

TILE-Q1: Tell us about your favourite project that you used tile in or another architect's work - interior, facade, etc.

A: Casa Vicens, which was the first house built by Antoni Gaudí. again an icon of catalan modernism where we find that the architect in this case really managed to take advantage of the properties of tile.

A: 카사 비센스(Casa Vicens)는 안토니 가우디가 지은 최초의 집이었다. 이 역시 카탈로니아 모더니즘의 아이콘으로, 건축가가 실제로 타일의 특성을 아주 잘 이용한 경우이다.

Casa Vicens ©jorapa

Casa Vicens ©TxllxT TxllxT

TILE-Q2: What are the strengths and weaknesses of tile?

A: Tile cannot work as a structure, but on the other hand the aggregation of tiles can be very beautiful worked in terms of patterns and geometries, thank to the multiple possibilities of colors and shapes that it can develop.

A: 타일은 구조로 쓸 수 없지만 타일의 집합은 패턴과 기하학 측면에서 매우 아름답다. 이는 개발할 수 있는 색상과 모양의 다양한 가능성 덕분이다.

GLASS-Q1: Tell us about your favourite project that you used glass in or another architect's work - interior, facade, etc.

A: The glass pavilion at the Toledo Museum of Art by SANAA. We find that it is a building that, by just using glass it makes us feel all the properties of the material.

A: SANAA가 지은 톨레도(Toledo) 미술관의 유리 파빌리온이다. 이 건물은 유리를 사용하는 것만으로 이 재료의 모든 특성을 느끼게 하는 건물이라고 생각한다.

GLASS-Q2: What are the strengths and weaknesses of glass?

A: The properties of transparency, flexibility, capability of changing of shape, make glass a unique material. Of course some weaknesses of it are its fragility, price and thermal behavior.

A: 투명성, 유연성과 모양을 변경할 수 있는 특성 덕분에 유리는 독특한 재료이다. 물론 단점은 부서지기 쉽다는 점과 가격 및 열적 거동이다.

WOOD-Q1: Tell us about your favourite project that you used wood in or another architect's work - interior, facade, etc.

A: The Wooden House by Sou Fujimoto. By just using one material in a certain shape (wooden prisms) and aggregating them in a certain way, he manages to create a very rich space.

A: 후지모토 소우의 목조 주택이다. 한 재료를 특정한 모양(나무 프리즘)으로 사용하고 특정한 방식으로 모음으로써 매우 풍부한 공간을 만든다.

WOOD-Q2: What are the strengths and weaknesses of wood?

A: Wood is a material that we use a lot, both in structures and façades, it is cheap and ecological, and has very good thermal properties.

A: 목재는 우리가 구조와 외관 모두에 많이 사용하는 재료이다. 값이 싸고 친환경적이며 열적 특성이 매우 좋다.

Toledo Glass Pavillion ©Adam C Nelson

TAKK Architecture 375

TheeAe Architects

Who is ...?

TheeAe is abbreviation of the evolved architectural eclectic. Its name is about the effort and dedication to the value of architectural aesthetic which shall be laid on place, history and culture of surrounding environment.

TheeAe pursues re-searching and re-finding the elements that have been embedded into the context of environment, so as to define the beauty of the architecture within the given context.

Q1: What is material to an architect (or to you)?

A: **Building's materiality is what we directly interact with through our physical senses.** We smell, touch and feel from it. In addition, it implies the mood through the texture and color tones. Moreover, it creates cultural characteristic through the indigenous local materials. Overall, the materiality is an inseparable element for the wholeness of architecture.

A: 건물의 물질성은 우리가 물리적 감각을 통해 직접 상호작용하는 것이다. 냄새를 맡고, 만지고, 느낄 수 있다. 또한, 질감과 색조를 통해 분위기를 암시한다. 더욱이, 지역의 고유 재료로 문화적 특성을 창출한다. 전반적으로, 물질성은 건축 전체에서 분리할 수 없는 요소이다.

Q2: Tell us about your favourite (or most often used) material and why.

A: We use metallic substance so much. It is not intentional but rather our choices are coming from the experience and familiarity on installation and its reliability on the quality. Especially, its flexible and versatile form on application for building façade or interior decoration is the main reason that we naturally used to employ them.

A: 우리는 금속 물질을 굉장히 많이 사용한다. 이는 의도적인 것이 아니라 설치에 대한 경험과 친숙함, 품질에 대한 신뢰성이 낳는 선택이다. 특히 외관이나 실내 장식을 위한 응용이 유연하고 형태가 다양하기 때문에 자연스럽게 금속을 사용하게 된다.

Q3: When do you decide the material during the design process and what is your criteria? (e.g. budget, client's preference, design concept, climate, etc.)

A: It is difficult to tell. We need to consider all possible conditions before we decide certain materials. Of course, when we start to design, normally materials may not be the first consideration to come up with. However, not always is it the case. Depends on the types of building, location, functions and many other reasons, material decision may vary to determine when and which.

Q4: What are some architectural projects that inspired you regarding brick, tile, wood and/or glass? And why?

A: Project that I got inspired is not just about the materials. It is holistic sense of design as wholeness of architecture. So very hard to answer on this question.

Q5: Tell us about the materials you are interested in or want to use in your projects right now.

A: Not haven't considered about it yet.

A: 말하기 어렵다. 우리는 특정 재료를 결정하기 전에 가능한 모든 조건을 고려해야 한다. 물론 우리가 설계를 시작할 때, 재료는 보통 첫 번째 고려 사항이 아닐 수도 있다. 하지만 항상 그런 것은 아니다. 건물의 종류, 위치, 기능 및 기타 여러 가지 이유에 따라 재료가 언제 어떻게 결정될지 다를 수 있다.

A: 내가 영감을 받는 프로젝트는 재료에 관한 것만이 아니다. 건축의 전체성으로서 디자인의 전체적인 느낌이 중요하다. 그래서 이 질문에 답하기 매우 어렵다.

A: 아직 생각해 본 적 없다.

BRICK-Q1: Tell us about your favorite project that you used brick in or another architect's work - interior, facade, etc.

A: There quite many projects we like. If we want to refer one of them, we would like to refer ABC building in South Korea designed by Wise Architecture. The use of brick was well utilized to bring the contrast of density through the gap of the space through the installation of bricks. Also the massing of the building is well expressing of this contrast. That is what we believe how architecture simply can be done well by the use of materials.

A: 우리가 좋아하는 프로젝트가 꽤 많다. 그중 하나를 꼽자면, 우리는 와이즈 건축(Wise Architecture)이 디자인한 한국의 ABC 건물을 언급하고 싶다. 여기서 벽돌은 공간의 간격을 통해 밀도의 대비를 일으키는 데 잘 사용되었다. 건물의 매싱 또한 이러한 대비를 잘 표현하고 있다. 이 프로젝트는 재료를 사용하여 건축이 어떻게 단순하게 잘 지어질 수 있는지 우리가 믿는 바를 보여준다.

BRICK-Q2: What are the strengths and weaknesses of brick?

A: For answering of advantage of brick, it is almost maintenance free material. No need for paintings, caulking, or staining for years and years. In addition, it protects sound from the outside wall, and it is fire resistant, looking very natural and comfortable, and most importantly environmental friendly.
For disadvantages, it can be expensive due to much involvement on labor. Thus, too

A: 벽돌의 장점은 관리가 거의 필요 없는 재료라는 점이다. 오랜 시간동안 페인트나 코킹, 스테인 같은 관리가 필요 없다. 또한, 벽돌은 외부로부터 소리를 차단하고, 내화성이 있고, 매우 자연스러우며 편안하지만, 가장 중요한 장점은 친환경적인 점이다. 벽돌의 단점은 건설하는데 필요한 노동량이 많아 비용이 많이 들 수 있다는 점이다. 그리

heavy to require strong foundation. Also, it takes time to build well.

고 매우 무겁기 때문에 튼튼한 토대가 필요하다. 또한 잘 지으려면 시간이 걸린다.

TILE-Q1: Tell us about your favourite project that you used tile in or another architect's work - interior, facade, etc.

A: Tile is most widely used interior materials in building industry. We use tile for floors and walls. Even tiles can be easily customized with any pattern or images. The floor pattern we designed for one of our interior project was baked in the factory and installed on floor. The process of ordering the product from the manufacturer was not simple, since we needed to make numerous times of reproduction due to unsatisfactory on color differences. However, overall quality

A: 타일은 건축 산업에서 가장 널리 사용되는 실내 자재이다. 우리는 바닥과 벽에 타일을 사용한다. 타일은 패턴이나 이미지에 맞춰 쉽게 맞춤제작까지 할 수 있다. 우리가 디자인한 인테리어 프로젝트의 바닥 패턴은 공장에서 구워서 바닥에 설치했다. 색상 차이가 만족스럽지 않아 여러 번 재생산해야 했기 때문에 제조업체에서 제품을 주문하는 과정은 간단하지 않았지만 설치 후 전반적인

The Artist House by Avoid Obvious + TheeAe

after installation was remarkable.

품질은 놀라웠다.

TILE-Q2: What are the strengths and weaknesses of tile?

A: Most of our interior projects were done with the use of (ceramic) tiles for the finish materials. There are truly tons of choices designers can choose from and it is also easy installation by semi-skilled cement workers. Also, its application is quite simple compare to other building material installation.

In contrast, the weakness of using tiles might its strength. It is easily broken when it is dropped on floor and it requires time and labour to install one by one as same as brick installation.

A: 우리 인테리어 프로젝트의 대부분은 마감재로 (세라믹) 타일을 사용했다. 디자이너가 고를 수 있는 선택의 폭이 정말 넓으며 적당히 숙련된 시멘트공이라면 쉽게 설치할 수 있다. 또한, 다른 건축 자재와 비교하여 설치가 매우 간단하다. 대조적으로, 타일을 사용하는 단점은 내구성이다. 바닥에 떨어졌을 때 쉽게 부서지며 벽돌과 마찬가지로 하나씩 설치하는 데 시간과 노동이 필요하다.

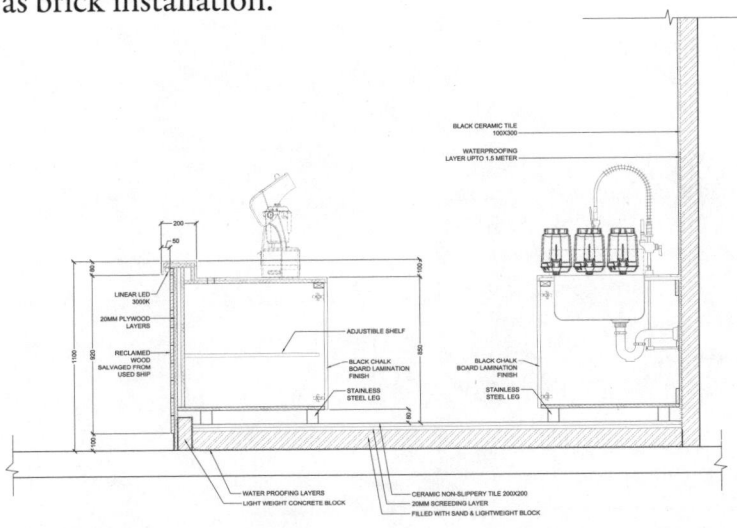

SECTION DETAIL

The Artist House by Avoid Obvious + TheeAe

TOUCH Architect

Who is ...?

TOUCH Architect Co.,Ltd. was first established since 2014. It was changed from TOUCH STUDIO Architect partnership, with four years experiences into a company. With a great chance of an improvement, we have two main co-founders consist of Mr. Setthakarn Yangderm as an architect and leader of our firm and Ms. Parpis Leelaniramol as an architect, which will corporate together in design, construction, and management.

Q1: What is material to an architect
(or to you)?

A: Building materials are part of an architecture, as well as creating various type of characteristics through its visibility and feeling of texture. It is an integration between visual quality and structural stability. Appropriate use of different kind of materials should be considered by usage and the context of it. **It is not only for function and practical to the building, but also provide the aesthetic to the architecture. It becomes part of architecture which indicates the timeline of an architecture evolution as well.**

A: 건설 자재는 건축 일부이다. 가시성과 질감을 통해 다양한 유형의 특성을 만들며 시각적 품질과 구조적 안정성의 통합이기도 하다. 여러 종류의 자재를 적절히 사용하려면 건물의 용도와 그 콘텍스트를 고려해야 한다. **자재는 또한 건물의 기능과 실용성뿐만 아니라 건축에 미적 감각을 부여하고 건축 진화의 연대를 나타내는 건축 일부가 된다.**

Q2: Tell us about your favourite
(or most often used) material and why.

A: Wood is our most favourable building material because of many reasons. First, it is physically strong, which is suitable for being a primary structure of the building, while also light and flexible compares to other structure such as concrete and steel. It has been being used as a structure since the ancient time before others. Secondly, it is a sustainable and environmental

A: 나무는 여러 이유로 우리가 가장 선호하는 건설 자재이다. 첫째, 물리적으로 튼튼하기 때문에 건물의 주요 구조로 적합하고 콘크리트 및 강철과 같은 다른 구조와 비교해 가볍고 유연하다. 다른 자재보다 먼저 고대부터 구조물을 짓는 데 사용되어 왔다. 둘째, 나무는 유일하게 재생할 수 있으며 지속 가능한 친환경적인 건축 자재

friendly construction material which is the only renewable one. In terms of cutting the trees to make wooden material seems damaging the world. On the other hand, in positive way, it can be replaced by cutting one, and planting more, which helps reduce climate change, as well as increase biodiversity. Moreover, wood is not being used for only structure, it is also used for finishing which convey harmonious feeling to the nature.

Q3: When do you decide the material during the design process and what is your criteria? (e.g. budget, client's preference, design concept, climate, etc.)

A: Initially, design criteria in choosing which material is the most suitable for each project, is being considered by the context. It should harmonize with its surrounding, as well as easy to find and easy to construct. Moreover, climate concern is also another factor of choosing, since Thailand's weather is hot and humid, thus, using an insulation wall materials might help reduce heat, while durable for humidity. The budget is the last step for the decision. We do not ignore this criterion by using only

이다. 나무를 잘라 자재를 만드는 것은 어떻게 보면 세상을 손상하는 것 같다. 반면에 긍정적으로, 나무는 하나를 자르면 더 심는 것으로 대체 될 수 있으며, 이는 기후 변화를 줄이고 생물 다양성을 증가시키는 데 도움이 된다. 더욱이 나무는 구조에만 사용되는 것이 아니라 자연과 조화로운 느낌을 주는 마무리에도 사용된다.

A: 각 프로젝트에 가장 적합한 재료를 선택하는 기준은 처음에 콘텍스트에 의해 좌우된다. 찾기 쉽고 건설하기 쉬울 뿐만 아니라 주변과 조화를 이루어야 한다. 더욱이 태국 날씨는 덥고 습하기 때문에 벽에 단열 재료를 사용하면 습도에 내구성이 있으면서도 열을 줄이는 데 도움이 될 수 있으므로 기후도 고려사항 중 하나이다. 결정의 마지막 단계는 예산이다. 우리는 아무 생각 없이 고가의 재료만을 사용하거나 해서 이 기준을 무시하지 않지만, 일단 가장 적합한 재료를

"WOOD IS NOT BEING USED FOR ONLY STRUCTURE, IT IS ALSO HELP YOU FEEL HARMONY WITH NATURE."

PLAA-GUT

high cost materials without any concern, but consider after the choosing most suitable material, and scope down to the appropriate one, since the same material has various price to choose.

먼저 선택한 후에 예산을 고려한다. 같은 재료라도 선택할 수 있는 가격의 폭이 다양하기 때문에 적절한 재료를 찾을 때까지 그 범위를 줄여나간다.

Q4: What are some architectural projects that inspired you regarding brick, tile, wood and/or glass? And why?

A: There are two architectural projects which we used to visit, inspiring us in material elements. The first one is, Alila Ubud Resort at Bali, Indonesia, which is designed by Kerry Hill Architects. The architect has made use of traditional Balinese design, but transforming it into modern geometry in the exotic creation. They designed a unique wall by creating different orientations and patterns which translate Balinese language through the pattern. It also plays with light and shadow by adding and subtracting the layer of bricks.

A: 우리가 자주 방문했던 건축물 중 자재에 대한 영감을 주었던 프로젝트는 두 개가 있다. 첫 번째는 케리 힐 건축사무소(Kerry Hill Architects)가 디자인한 인도네시아 발리의 알리어 우붓 리조트(Alila Ubud Resort)이다. 여기서 건축가는 전통적인 발리 디자인을 사용했지만 현대 기하학으로 변형해 이국적인 창조물을 만들어냈다. 발리 언어를 번역하는 다양한 방향과 패턴을 만들어 독특한 벽을 디자인했다. 또한 벽돌 층을 더하고 빼서 빛과 그림자를 여러모로 활용했다.

Alila Ubud Resort at Bali

The second one is 178 PRADA AOYAMA, a Prada store in Tokyo, designed by Swiss architects, Herzog & de Meuron. It is the rhomboid-shaped grid structure with an extraordinary glass façade. This building is not only using glass just for displaying products inside, but convex and concave glass is also generates facetted reflections, which enable viewers from both outside and inside the building to see the different effects of the inside products and city outside. It is noticeable and serve all functions needed.

두 번째는 스위스 건축가 헤르조그 앤 드 뫼롱(Herzog and de Meuron)이 디자인한 도쿄의 프라다 매장인 178 프라다 아오야마 점이다. 이 건물은 보기 드문 유리 외관을 지닌 마름모꼴 모양의 격자 구조이다. 이 건물은 내부 제품을 전시하기 위해 유리를 사용할 뿐만 아니라 볼록하고 오목한 유리로 건물 외부와 내부의 사용자 모두가 내부에 있는 제품과 외부에 있는 도시를 다른 느낌으로 볼 수 있는 다면의 반사를 생성한다. 이 건물은 눈에 잘 띄면서도 필요한 모든 기능을 수행한다.

Q5: Tell us about the materials you are interested in or want to use in your projects right now.

A: Bamboo is a very interesting materials but hardly used in our projects. It is a local material in Thailand which is easily to find, while having their own aesthetic. It can be used for all part of the building, from the structure to the finishing or even the small elements. Moreover, it creates a tropical sensation which is suitable for country home and resort even if it is not as durable as other materials, but it is easily be replaced.

A: 대나무는 매우 흥미로운 재료이지만 우리 프로젝트에는 거의 사용하지 않는다. 태국의 현지 자재로 쉽게 찾을 수 있으며 그만의 미적 감각이 있다. 건물의 모든 부분, 구조에서 마무리 또는 다른 세세한 곳까지 사용할 수 있다. 더욱이 내구성이 좋지 않아도 시골집과 휴양지에 적합한 열대성 느낌을 줄 수 있다. 하지만 다른 재료로 쉽게 대체될 수 있다.

178 PRADA AOYAMA

BRICK-Q1: Tell us about your favorite project that you used brick in or another architect's work - interior, facade, etc.

A: "The Rest Avenue" is a neighborhood center in Surin, which is located at North-Eastern part of Thailand, well-known as an ancient city with Prasart; a load-bearing wall structure castle which using all stone-brick. This place consists of food and souvenir retail shops originated from characteristic of Surin.
The 3 elements from 'Prasart' which are PRASART's LAYOUT, BRICK PATTERN, and GRID SYSTEM are being used, and combine them together in order to created an overall plan. The 4 main entrances also have their features related to Surin which are;
NORTH | An Elephant gate
EAST | A Rice gate
SOUTH | A Silk gate
WEST | A Khmer gate.
This West gate is the one that use all brick pattern transmitted from the ancient pattern of stone-brick masonry. It integrates modern pattern with an ancient character.

A: 레스트 애브뉴(The Rest Avenue)는 쁘라삿이 있는 고대 도시로 잘 알려졌고 태국의 북동부에 위치한 수린의 이웃 센터이다. 쁘라삿은 돌과 벽돌을 사용하는 하중 지지 벽 구조의 성이다. 레스트 애브뉴는 수린의 특색을 살린 음식점과 기념품 가게로 이루어져 있다.

쁘라삿의 세가지 특징인 배치, 벽돌 패턴과 그리드 시스템을 사용했고 전반적인 평면을 위해 이들을 결합했다. 4개의 주요 입구에도 수린과 관련된 특징이 있다.
북쪽 입구-코끼리 문
동쪽 입구-쌀 문
남쪽 입구-비단 문
서쪽 입구-크메르 문

이 서쪽 문은 석공을 쓰는 고대 패턴에서 전해지는 벽돌 패턴을 사용한 문이며 현대적 패턴과 고대의 특징을 통합했다.

The Rest Avenue

BRICK-Q2: What are the strengths and weaknesses of brick?

A: Brick material, in general, is durable and highly fire resistant. Since it is a very strong material, that can be used as a load-bearing wall and helps to support the

A: 벽돌 재료는 일반적으로 내구성이 뛰어나고 내화성이 높다. 매우 튼튼한 재료이기 때문에 하중 지지 벽으로 사

building structure. Moreover, it is easy to construct which does not require any finishing or painting if not necessary, while still have their own aesthetic by creating different orientations and patterns. However, it does have some weakness. Heavy weight will increase stress to the piling and foundation which is needed to support the bricks. Moreover, it consumes more time to create such a wall.

용할 수 있으며 건물 구조를 지지하는데 도움이 된다. 더욱이, 디자인에 필요하지 않으면 특별히 따로 마감이나 페인팅이 필요하지 않다. 다양한 방향과 패턴을 만들 수 있기 때문에 그만의 미학도 있다. 하지만 벽돌에도 약점은 있다. 무게가 무겁기 때문에 그를 지탱하는데 필요한 말뚝과 건물의 기초에 부담이 늘어난다. 더욱이 그런 벽을 만드는 데 더 많은 시간이 소모된다.

TILE-Q1: Tell us about your favourite project that you used tile in or another architect's work - interior, facade, etc.

A: Sukpisan SHRINE is a 'Brahma shrine' combining with a 'house of Guardian spirit'. Since the overall context is modern contemporary architecture, the shrine was designed by following the context which finally becomes a modern-like shrine but still keep traditional function as it should be.
The architecture has 3 gables, big one, medium one, and small one which is for MAIN ENTRANCE, BRAHMA, and GUARDIAN SPIRIT respectively. Floor level has been elevated for 1 meter

A: 숙피산 신사는 수호신의 집과 결합된 브라마*의 신사이다. 전반적인 콘텍스트는 현대 건축이기 때문에 신사는 그에 따라 설계됐고 마침내 현대적이지만 여전히 전통적인 기능을 유지하는 신사가 되었다. 이 신사에는 정문, 브라마, 수호신용으로 각각 대,중,소의 박공지붕이 있다. 신사는 집보다 높은 곳에 배치해야 하므로 신사의 바닥이 집보다 1 미터 높으며 쉽게 접근하고 예배할 수 있는 각 건물 사이의 큰 녹지에 위치한다. 게다가 브라마는 4개의 얼굴을 가

because the shrine should be placed on higher level than the house. It is located in a large green space in between each building which can easily access and worship. Moreover, Brahma has 4 faces, in our belief, 4 faces can see things and protect us in 360 degrees.

The big gable for main entrance was designed to make the people who walking up to the shrine, should 'BOW THEIR HEADS' since the gable roof has been elevated down at the last step of staircase, in order to force everyone to salute the Brahma.

Material using of this shrine is mainly 'white marble tile' cladding on reinforced concrete structure. It has been used in all floor, wall, and roof, since the continuity of massing was created. Moreover, tile material and lighting design shows different effects between DAYTIME and NIGHTTIME. During daytime, golden Brahma can easily be seen from 'white marble tile' background, while 'recessed lighting' makes the golden Brahma be seen in darkness. The shrine shines and becomes a lightness during nighttime.

지고 있고 그 얼굴은 360도에서 사물을 보고 우리를 보호할 수 있다고 믿는다.

신사에 오는 모든 사람이 브라마에게 경의를 표하도록 하기 위해서 정문의 큰 박공지붕은 일정한 높이로, 마지막 계단에 오르면 머리를 숙이게 디자인되었다.

이 신사는 주로 철근 콘크리트 구조에 흰색 대리석 타일로 지었다. 용량감의 연속성을 위해 모든 바닥, 벽 및 지붕에 타일을 사용했다. 또한 타일 및 조명 디자인은 낮과 밤 간의 다른 효과를 만든다. 낮에는 황금 브라마가 흰색 대리석 타일 배경으로 눈에 띄며 밤에는 매입형 조명으로 황금 브라마를 어둠 속에서도 잘 볼 수 있다. 숙피산 신사는 밤에 반짝이고 빛이 된다.

*역주 – 브라마: 힌두교의 창조신

TILE-Q2: What are the strengths and weaknesses of tile?

"MATERIAL IS NOT ONLY FOR FUNCTION AND PRACTICAL TO THE BUILDING, BUT ALSO PROVIDE THE AESTHETIC TO THE ARCHITECTURE."

Sukpisan SHRINE

A: The tile itself, has a huge variety of sizes, shapes, colours, patterns, and textures. It can be used for flooring, wall cladding, and for some furniture such as, a counter top for kitchen. It creates different types of an architectural character through its colour and pattern, since it is a simulation form of natural elements, such as wood, concrete, granite, and marble. Besides, it helps protect the concrete floor and wall from cracking and dirt since it is easy to clean, as well as, being respect from most clients because of trusting in its quality with reasonable price. However, there is some cons using the tile. Even if the tile is not affected by humidity, it may cause some mold on tile grout which is not easy to be replaced.

A: 타일은 크기, 모양, 색상, 패턴 및 질감이 아주 다양하며 바닥재, 벽 혹은 주방 카운터와 같은 일부 가구에 사용할 수 있다. 목재, 콘크리트, 화강암이나 대리석과 같이 자연을 가장하기 때문에 색상과 패턴을 통해 다양한 유형의 건축적 특징을 만든다. 게다가 콘크리트 바닥과 벽을 균열과 먼지로부터 보호하는데 도움이 되며 청소하기 쉽다. 의뢰인 대부분은 타일이 합리적인 가격으로 품질을 제공하는 것을 믿고 쓴다. 하지만 타일을 사용하는데 몇 가지 단점이 있다. 타일이 습도에 영향을 받지 않더라도 타일 그라우트에 곰팡이를 약간 일으킬 수 있고 그라우트는 교체하기 쉽지 않다.

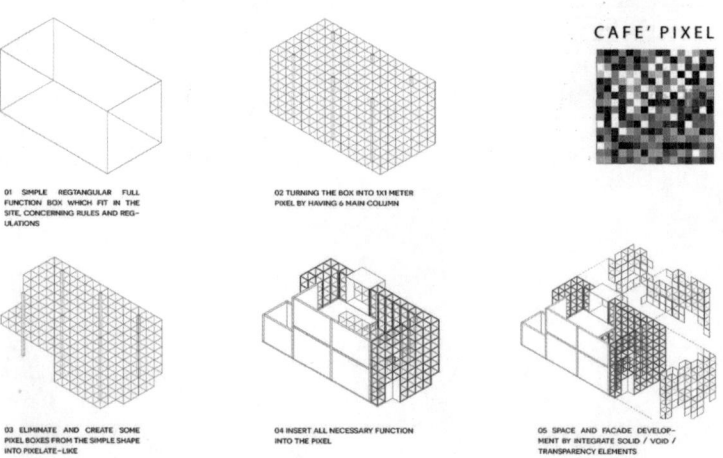

GLASS-Q1: Tell us about your favourite project that you used glass in or another architect's work - interior, facade, etc.

A: Café Pixel is located at the city center of Udonthani, Thailand. It has a limited land area which can be approached from main street directly. There are three primary functions which are café, restaurant, and bar, together in one space with time-sharing functions; café / bakery and healthy food restaurant in the morning, while using as a craft-beer bar in the evening.

Since it is a time-sharing, massing and zoning design of this café was created in the way to suit these both contrast functions. There are two floors, the first floor with café and bar counter, and the mezzanine floor for only customers' seats,

A: 카페 픽셀은 태국 우돈타니의 도심에 있으며 대로에서 바로 접근할 수 있는 토지 면적이 제한되어 있다. 이 한 공간에서 카페, 레스토랑과 바가 시간별로 동시 운영된다. 아침에는 카페/빵집 겸 건강한 음식을 파는 레스토랑이며 저녁에는 수제 맥주 바가 된다. 건물의 용도가 시간별로 나뉘므로 이 카페의 용량감 및 평면은 이 두 가지 대조되는 기능에 맞게 만들어졌다. 카페 픽셀은 층이 두 개가 있는데 1층에는 카페와 바 카운터가 있고 중이층에는 고객을 위한 좌석만 있다. 이 모두 건물 법규, 규칙 및 규정에 맞는 직사각형 상자에 맞게 디자인됐

Café Pixel

TOUCH Architect 397

which fit in a rectangular function box, by concerning building codes, rules, and regulations. In order to maximize the amount of seats, an open-plan and flexible seating is needed.

One simple box was turning to a pixelate box with 1x1 meter of each pixel. It creates a façade pattern, continuing into inside space. The pixel is not only used for façade, but also integrated and used as 'pixel bar seats' and a decoration shelf for displaying beverages items.

Material of the pixel is all glass, which contains both transparent clear glass and translucent glass, in order to portray different effects through each type of material and each amount of glass layers; single or double. White translucent glass will help reduce and reflect direct sunlight during daytime for café, while lighting design inside will glow as a craft-beer bar during nighttime. It creates different atmosphere for different time.

GLASS-Q2: What are the strengths and weaknesses of glass?

A: Attribute of glass is mainly transparency and translucence. It is the only one building material which allows

으며 좌석의 양을 극대화하기 위해서 오픈 플랜과 유연한 좌석 배치가 필요했다. 단순한 상자 하나가 1x1미터의 픽셀로 채워진 픽셀 상자로 바뀐다. 이는 파사드 패턴을 만들고 내부 공간으로 이어진다. 픽셀은 파사드뿐만 아니라 픽셀 바의 좌석 및 음료 품목을 장식할 수 있는 선반으로 사용된다.

픽셀의 재료는 전부 유리이다. 각기 다른 종류와 단일 혹은 이중 유리를 통해 다양한 효과를 나타내기 위해서 투명과 반투명 유리를 모두 썼다. 흰색 반투명 유리는 낮에 카페에 들어오는 직사광선을 줄이고 반사하는데 도움이 되며, 밤에는 조명이 수제 맥주 바를 빛낸다. 카페 픽셀은 다른 시간대에 따라 다른 분위기를 만들어 낸다.

A: 유리의 주된 속성은 투명성과 반투명성이며 빛이 지나가는 것을 허용하는 유일

the light passing by. It is not only creating an aesthetic to the façade, but also has been used for connecting indoor and outdoor space together, as well as allows for natural sunlight to come in, which consume less of indoor lighting electricity. However, using all glass wall is not suitable for hot climate since it absorbs heat and act as greenhouse effect which increase energy consumption by using more air condition. Nor, it is easily break which is not well for security condition.

On the other hand, there is some types of glass which solves the above mentioned disadvantages, for instance, Low-E glazing is a high thermal insulated glass which allows only natural sunlight, not the heat, and, there is a safety glass which cannot be broken. However, these further features come with higher cost.

한 건축 자재이다. 파사드에 미학을 만들어 낼뿐만 아니라 실내와 실외 공간을 연결하는 데 사용되며 자연 채광 덕분에 실내조명에 전기를 덜 소비할 수 있다. 그러나 유리는 열을 흡수하고 온실 효과를 일으킨다. 이 때문에 더 많은 에어컨을 사용하게 되고 에너지 소비를 증가시키기 때문에 모든 벽에 유리를 사용하는 것은 더운 기후에 적합하지 않다. 또한 쉽게 깨지기 때문에 보안을 위해서도 그다지 좋은 재료는 아니다.

반면에 위에 언급한 단점을 해결하는 유리 종류도 있다. 예를 들어, 열을 제외한 자연광만을 허용하는 저방사 유리로 로이유리(Low-E)가 있으며 깨지지 않는 안전유리도 있다. 그러나 이러한 부가 기능은 비용이 많이 든다.

WOOD-Q1: Tell us about your favourite project that you used wood in or another architect's work - interior, facade, etc.

A: It is our project names, Chan'4 ECO-STAY, which is located at Chantaburi, Thailand. It derived from characteristics of the site itself. Chantaburi province

A: 태국의 찬타부리에 위치한 이 프로젝트 이름은 찬 포 에 코 스테이(Chan'4 ECO-STAY) 이다. 이 건물은 사이트 자체의 특성에서 파생되었다. 찬

contains various types of geography, especially, sea and hill are the main typography of this location. This site does not only have sea and hill, but also has a swamp/ lake and mangrove inside. This homestay contains both accommodation and an organic farming, which is the owner's activity doing agriculture after retirement. So, this farm becomes a part of people who stay here that everyone can live and learn with ecology.

There are three levels which integrated the homestay within the house. Main approach is leading to lobby space at the ground floor that keeps characteristics of Thai traditional house which has an open space with flow circulation on the first floor, called "Tai-Toon". On the second floor, each rooms are suitable for family which can be sighted to the lake and sea. Moreover, there is common space on the roof top for multi-purpose activities for all.

Material using of the overall building is wood. It has been used from floors, walls, roof structure, to all façade elements, except the primary structure, column and beam, which is made of reinforced concrete. The reason is, using all wood in stead of steel can help avoiding extremely rust which occurs because of the sea

타부리 지방에는 다양한 유형의 지형이 있으며, 특히 바다와 언덕이 주요 지형이다. 이 사이트에는 주변에 바다와 언덕이 있을 뿐만 아니라 내부에 늪과 호수, 맹그로브가 있다. 이 홈스테이에는 숙박 시설과 유기농 농장이 있는데 이 농장은 의뢰인이 은퇴 후 하는 일이다. 이 농장에서는 모든 사람이 환경과 어우러져 살며 배울 수 있고, 지내고 가는 사람들의 일부가 된다. 이 건물에는 홈스테이를 통합하는 세 층이 있다. 주 진입로는 1층에 있는 로비 공간으로 이어진다. 이 로비는 1층에 동선이 흐르는 열린 공간이 있는 태국 전통 주택의 특성인 '타이 뚱(Tai-Toon)'을 유지한다. 2층에는 각 방이 가족 여행객에게도 적합하며 호수와 바다를 볼 수 있다. 또한, 지붕에는 모든 사람이 쓸 수 있는 다목적 활동을 위한 공용 공간이 있다.

건물 전체에 사용된 재료는 나무이다. 철근 콘크리트로 만들어진 기본 구조, 기둥 및 들보를 제외한 모든 파사드 요소, 바닥, 벽, 지붕 구조에서 사용했다. 그 이유는 강철 대신에 목재를 사용하면 바다 소금 때문에 발생하는 극도의 녹을 피하는 데 도움이 되기 때문이다. 또한, 아름다움과 기능적인 면에서 봐도, 조

salt. In addition, in terms of beauty and function, an adjustable wooden trellis façade allows for clear view and natural ventilation and harmonize with nature.

정 할 수 있는 나무 격자 파사드로 훤히 트인 전망이 보이고 자연 환기를 할 수 있으며 건물이 자연과 조화를 이룸을 알 수 있다.

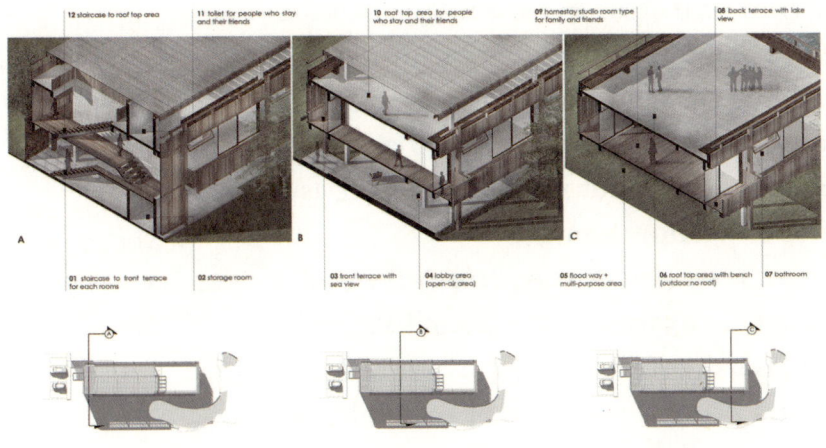

Chan'4 ECO-STAY

WOOD-Q2: What are the strengths and weaknesses of wood?

A: It cannot be denied that every material has their own weaknesses, wood either. Only hard wood can be used for structure which is much more expensive and hard to find, while soft wood can only be used for interior finishing, furniture, and cladding, which is not termite resistance. Some of them are easily to stretch and retract which is not suitable for using as a door or window frame.

Apart from weakness, tons of wood's strengths are already mentioned above, it is physically strong, which is suitable for being a primary structure of the building, while also light and flexible compares to other structure such as concrete and steel. It is yet sustainable and environmental friendly construction material which is the only renewable one. Moreover, wood is not being used for only structure, it is also used for finishing which convey harmonious feeling to the nature. It looks soft and comfy with style.

A: 모든 물질에는 나름의 약점이 있다는 것을 부인할 수 없으며 나무도 마찬가지이다. 견재(hard wood) 만이 구조에 사용될 수 있지만 비싸고 찾기 어려우며, 연재(soft wood)는 내부 마무리, 가구 및 클래딩에만 사용할 수 있으며 흰개미에 저항력이 없다. 연재 중 일부는 늘어나거나 수축하기 쉬워서 문이나 창틀로 사용하기에 적합하지 않다. 단점 외에도, 수많은 나무의 장점은 이미 위에 언급했다. 물리적으로 강하여 건물의 기본 구조에 적합하고 콘크리트 및 강철과 같은 다른 구조와 비교해 가볍고 유연하다. 지속할 수 있고 환경친화적인 건축 자재이며 유일하게 재생할 수 있다. 또한 나무는 구조에만 사용되지 않고 자연과 조화로운 느낌을 전달하는 마무리에도 사용된다. 부드러워 보이고 스타일 있게 편해 보이기도 한다.

UNStudio

© Inga Powilleit

Who is ...?

Ben van Berkel studied architecture at the Rietveld Academy in Amsterdam and at the Architectural Association in London, receiving the AA Diploma with Honours in 1987.

With UNStudio he realised amongst others the Mercedes-Benz Museum in Stuttgart, Arnhem central Station in the Netherlands, the Raffles City mixed-use development in Hangzhou, the Canaletto Tower in London, a private villa up-state New York and the Singapore University of Technology and Design.

Q1: What is material to an architect (or to you)?

A: Materials are the tools we use to organise life. Architect's however need to play and experiment with materials, not only as finishes, but also for the structural elements of their architecture.

A: 재료는 우리가 생명을 구성하는 도구이다. 그러나 건축가는 마무리만이 아니라 건축의 구조를 위해 재료를 여러모로 활용하며 실험해봐야 한다.

纹路变化 STONE TEXTURES

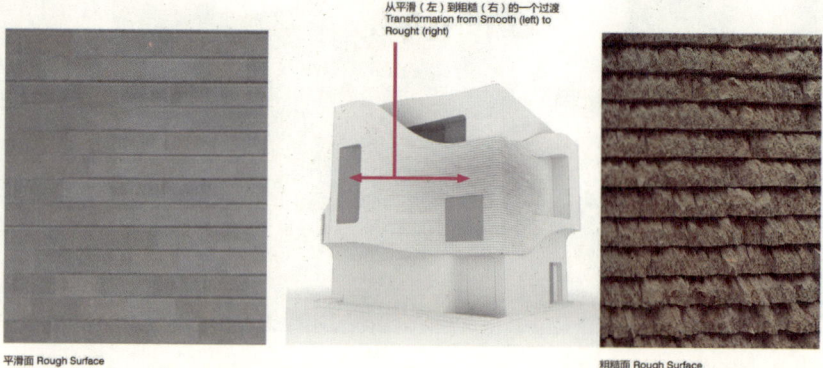

平滑面 Rough Surface

从平滑（左）到粗糙（右）的一个过渡
Transformation from Smooth (left) to Rought (right)

粗糙面 Rough Surface

Q2: Tell us about your favourite (or most often used) material and why.

A: I am particularly fascinated by Basalt lava stone as a material. This stone embodies quickly solidified movement, heat and energy, but is also extremely strong. Glass is also becoming a more versatile material: it is now possible to produce glass with double curves for

A: 나는 현무암 용암석에 특히 매료되었다. 이 돌은 빠르게 응고된 움직임, 열 및 에너지를 구현하면서 매우 견고하다. 유리도 보다 다용도가 되고 있다. 예를 들자면 이중 곡선이 있는 유리를 생산할 수 있으며, 계단, 벽 및 기둥과 같

instance and you increasingly see glass being used in stairs, walls and even for structural elements, like columns. Mostly I am interested in natural materials that can also be used for engineering or in the structure of buildings.

Q3: When do you decide the material during the design process and what is your criteria? (e.g. budget, client's preference, design concept, climate, etc.)

A: **Most important to me is that the material you use has to fit the concept and the context of the project, in both a literal and a metaphorical sense.** But as you mention, climate and budget also play an important role. Our choice of the final material is always set up in a dialogue with the client, where usually we present a number of possible options and together decide on which is the most appropriate.

Q4: What are some architectural projects that inspired you regarding brick, tile, wood and/or glass? And why?

A: For me Gaudi was the best tile architect we have known to date, while

406 Brick, Brick! What do you want to be?

The main square of park Güell © Isiwal

with brick work, Michel de Klerk from the Amsterdam School was able to produce the most wonderful and complicated curves.

이다. 벽돌 작업으로는 암스테르담 학파의 미셸 드 클레르크(Michel de Klerk)가 가장 훌륭하고 복잡한 곡선을 만들었다.

Apartment building Het Schip in Amsterdam © Jvhertum

Q5: Tell us about the materials you are interested in or want to use in your projects right now.

A: Wood has become a very fascinating material for me, as we can now source it from replenishable forests and it is 100% recyclable. It this way it is a completely circular material in terms of sustainability.

A: 나무는 이제 보충 가능한 숲에서 얻을 수 있고 100% 재활용할 수 있기 때문에 나에게 매우 매혹적인 재료이다. 이렇게 지속 가능성 측면에서 보면 완벽하게 순환하는 재료이다.

Apartment building Het Schip in Amsterdam © Jvhertum

BRICK-Q1: Tell us about your favorite project that you used brick in or another architect's work - interior, facade, etc.

A: For the Villa Wilbrink single family house that we designed in the early 1990s, we used a technique for the facade where we glued the bricks together, instead of using cement. Because the layer of glue was so thin, the bricks were much more closely stacked together and it gave a really wonderful, solid effect to the house.

A: 1990년대 초에 우리가 설계한 빌라 윌브링크(Villa Wilbrink) 단독 주택을 위해 우리는 파사드에 시멘트 대신 풀로 벽돌을 붙이는 기법을 사용했다. 풀이 매우 얇았기 때문에 벽돌이 훨씬 더 밀착되어, 집에 정말 훌륭하고 견고한 효과를 낳았다.

BRICK-Q2: What are the strengths and weaknesses of brick?

A: Brick laying of course doesn't require huge machinery, it can be done by hand, which makes it great for smaller scale projects. However, brickwork is not strong enough to be used for very large spans, unless the walls are very thick - which is why we don't ever see brick skyscrapers.

A: 벽돌을 놓는 일에는 물론 거대한 기계가 필요하지 않고, 손으로 할 수 있기 때문에 소규모 프로젝트에 적합하다. 그러나 벽이 매우 두껍지 않은 한, 벽돌은 매우 큰 경간에 사용할만큼 강하지 않다. 그래서 벽돌로 된 고층 건물이 없는 것이다.

TILE-Q1: Tell us about your favourite project that you used tile in or another architect's work - interior, facade, etc.

A: The was Tower in Dubai I think is the most exciting use of tile that we have developed to date.
In this project the specially designed tiles principally have sustainable functions, whereby they lessen the heat load on the building. But they also give the facade an unusual visual effect, which relates to the scale of product design.

A: 두바이의 탑(Tower in Dubai)은 지금까지 우리가 개발한 가장 흥미로운 타일 사용법이라고 생각한다. 이 프로젝트에서 특별히 디자인된 타일은 주로 건물의 열부하를 줄이는 지속 가능한 기능을 한다. 거기에 파사드에 특이한 시각 효과를 더하며 이는 제품 디자인의 규모와 관련된다.

TILE-Q2: What are the strengths and weaknesses of tile?

A: Their lack of strength is literally their weakness. Although the strength is

A: 힘의 부족은 말 그대로 타일의 약점이다. 비록 내구력

Wasl Tower-Ceramic solar shading system

UNStudio 411

improving all the time, we still need to be careful with sizes, as these materials can break very easily.

이 항상 향상되고 있지만, 이런 재료는 매우 쉽게 깨지기 때문에 여전히 크기에 주의해야 한다.

GLASS-Q1: Tell us about your favourite project that you used glass in or another architect's work - interior, facade, etc.

A: In the La Defense offices in Almere (NL) the facades are clad with glass panels in which a multi-coloured foil is integrated and, depending on the time of day and the angle at which you view them, a variety of different colours are reflected. The office workers there have said that the effect is as if the sun shines in that courtyard every day of the year.

A: 네덜란드 알미르(Almere)에 있는 라데팡스(La Defense) 사무용 건물의 파사드에 여러 가지 색상의 호일이 통합 된 유리 패널을 썼다. 시간 및 시각에 따라 다양한 색상이 반사된다. 건물에서 근무하는 사람들은 마치 일년 내내 매일 안뜰에서 태양이 빛나는 것 같다고 말했다.

GLASS-Q2: What are the strengths and weaknesses of glass?

A: Philip Johnson's Glass House has already proved that a whole house can be built from glass, so strength is no longer the issue it once was. However too much glass also raises questions regarding sustainable heat loads.

A: 필립 존슨 (Phillip Johnson)의 글래스 하우스(Glass House)가 집 전체를 유리로 지을 수 있다는 것을 이미 증명했기 때문에 내구력은 더 이상 예전처럼 문제가 아니다. 그러나 너무 많은 유리는 지속 가능한 열 부하에 관한 의문을 제기한다.

WOOD-Q1: Tell us about your favourite project that you used wood in or another architect's work - interior, facade, etc.

A: We used wood in the facade cladding of a single family house that we completed a few years ago, called the W.I.N.D. House. The wood is hydro-thermic treated, which results in a long durability, while the wood still appears untreated and natural. In time, it changes to a grey colour, which refers to the tones of driftwood found on a nearby beach and the dunes that surround the house.

A: 몇 년 전에 완성한 W.I.N.D 하우스 (W.I.N.D House)라는 한 단독주택의 파사드에 나무를 사용했다. 목재를 수열처리하여 내구성을 높였고 화학처리되지 않아 자연적으로 보인다. 시간이 지남에 따라 목재는 회색으로 바뀌며 이는 인근 해변에서 볼 수 있는 유목과 집을 둘러싸고 있는 모래 언덕을 나타낸다.

WOOD-Q2: What are the strengths and weaknesses of wood?

A: People often worry that wood will not last long enough, or will require too much maintenance, but in fact today there are treatments that can make wood very durable.

A: 사람들은 종종 나무가 충분히 오래 지속되지 않거나 유지를 위해 너무 많은 보수가 필요하다고 걱정하지만 실제로 오늘날에는 나무의 내구성을 매우 높일 수 있는 처리법이 있다.

"MATERIALS ARE THE TOOLS WE USE TO ORGANISE LIFE."

W.I.N.D. House ©Fedde de Weert